THE UNDERTAKER AT WORK:
1900–1950

Brian Parsons has worked in the funer   stry in London since 1982. His PhD focused on the impact   e funeral industry during the twentieth century. I   ntributing to academic and trade journal   *1 Way of Death* (2001 Sutton Publishi   *me: The Development of Cremation in Nir*   2005 Spire Books), the history of two London fur   .1s, and (with Hugh Meller) the fourth edition of *London*   *2: An Illustrated Guide and Gazetteer* (2008 Sutton Publishing).

THE UNDERTAKER AT WORK: 1900–1950

BRIAN PARSONS

▲ An undertaker leads a coffin draped with a black pall being carried on a frame shoulder-high along King Street in Twickenham around the time of WWI

Strange Attractor Press
MMXIV

First published by Strange Attractor Press 2014
©Brian Parsons 2014

First edition 2014
ISBN: 978-1-907222-28-3
Brian Parsons has asserted his moral right to be identified as the author of
this work in accordance with the Copyright, Designs and Patents Act, 1988.
All rights reserved. No part of this publication may be reproduced in any
form or by any means without the written permission of the publishers.
A CIP catalogue record for this book is available from the British Library.

Design and layout: www.emeraldmosley.com

Strange Attractor Press
BM SAP, London, WC1N 3XX, UK
www.strangeattractor.co.uk
Printed in China

Dedicated to Bunny France who started work with A France &
Son of Holborn during the period covered by this book and is
still an undertaker at work.

CONTENTS

INTRODUCTION

The Undertaker at Work: 1900–1950 gives a glimpse into a period of considerable change and many challenges for funeral service. The book is in two parts. The 11 chapters comprising the first part provide a chronological and sequential account of significant events that have impacted upon funeral service during the first half of the twentieth century. Chapters 1 and 2 chart the introduction of embalming in 1900 and the formation of the British Embalmers' Society (BES). The British Undertakers' Association (BUA) emerged in 1905 out of the BES and was followed by other organisations such as the British Institute of Embalmers (BIE). The final part of chapter 2 looks at the BUA's transition into the National Association of Funeral Directors (NAFD). Following the Salisbury rail crash of 1906, the work of the embalmer is explored in chapter 3, while issues confronting funeral service during WWI and the 1918 Spanish flu pandemic are examined in chapters 4 and 5. With the BUA becoming a trade association in 1917, chapter 6 outlines the difficulties one undertaker encountered when returning from the war to establish his business. Around the same period, Co-operative societies started to establish funeral services and this structural change to the industry is discussed in chapter 7. The two chapters following this reveal how undertakers were involved in two contrasting cases of transportation of the dead, while chapter 10 continues this theme by assessing the work of undertakers after the R101 airship crashed in France and bodies were returned to England for burial. The final chapter looks at how funeral directors coped with restricted supplies and also the death of civilians during WWII. With three exceptions, all chapters were originally published as articles; in preparation for publication in this book, they have been extended and updated with new material.

The second part is a collection of images showing the undertaker at work between 1900 and 1950. Whilst funeral processions depict the most visible side of funeral service, images of premises, transport and behind- the-scenes activities such as coffin construction and embalming are also included. Drawn from a wide range of sources, many have never been published before.

This book is a companion to *From Undertaker to Funeral Director*, which focuses on organisational change in the funeral industry and the professionalisation of the funeral director. Although there is some overlap, this has been kept to a minimum as the chapters give greater background to historic events only mentioned in *From Undertaker to Funeral Director*.

In an age when 'funeral director' is the preferred occupational description, an explanation about the title of this book is necessary. The term 'undertaker' has been used throughout history to describe the person who undertook or accepted responsibility to bury the dead: when, in 1905, the BUA was founded, this term was used in the association's title. However, the need for a new descriptive term was prompted through the increasing custody for the dead, in addition to the growing complexity of funerals. In recognising these developments, the BUA became the National Association of Funeral Directors in 1935. However, as most of this book embraces a period when 'undertaker' was in frequent use, it was felt appropriate to include it in the title and in the text of all chapters except the last.

Brian Parsons
London 2014

ACKNOWLEDGEMENTS

I am very grateful to those who have helped in the preparation of this book. Many contributed to the original research upon which eight of the chapters are based, whilst others have given information about images. In particular I would like to thank Dr Philip Smyth (JH Kenyon), Christopher Henley (for access to material relating to his late father, Des Henley FBIE OBE), John Clarke, Peter Wilson, Dr Ian Hussein, Barry Albin (FA Albin & Sons), James Blackburn, Jim Gilbey, Mary Byrne, Reina James, Michael Ball (National Army Museum), Peter and Andrew Miller (WG Miller), Michael and Bunny France (A France & Sons), Stephen White, the Revd Dr Peter Jupp, Dr Tony Trowles (Westminster Abbey), Michael and Ann Nodes (John Nodes & Sons), Muriel Ghys, Drawn Trigg (FW Paine), Sandy Brophy, Peter Collins (Imperial War Museum), David Fletcher (Tank Museum), Andy Robertshaw (Royal Logistics Corps Museum), David Hall (Vintage Lorry Funerals), Sarah Burgess, Toni Slade (East London Cemetery), Dr Julian Litten, Richard Meunier, Dean Reader, Sandra Mitchell and Belinda Jones.

Staff at the following London local studies libraries: Newham, Hackney, Brent, Haringey, Tower Hamlets, Islington, Croydon, Camden, Greenwich, Lewisham, Southwark, Enfield, Richmond upon Thames, Royal Borough of Kingston and the City of Westminster.

Staff at London Metropolitan Archive, National Archive and the London Transport Museum. Staff at the following funeral directors: Wm Denys, J Jeffries, CE Hitchcock, JH Kenyon and FW Paine.

I am grateful to organisations and individuals who have given permission for images and other material to be reproduced. In particular, I would like to thank the National Association of Funeral Directors (*BUA Monthly* and *The National Funeral Director*) and Heritage Studios Ltd (*The Undertakers' Journal* and *Funeral Service Journal*) to reproduce images from their publications. Where a source for an image is not stated, it is held in the author's collection.

ABBREVIATIONS

Publications

FSJ	*Funeral Service Journal*
TUJ	*The Undertakers' Journal*
TUFDJ	*The Undertakers' and Funeral Directors' Journal*
BUA Monthly	*British Undertakers' Association Monthly*
TNFD	*The National Funeral Director*

Trade Associations

BES	British Embalmers' Society
BFWA	British Funeral Workers' Association
BIE	British Institute of Embalmers
BIU	British Institute of Undertakers
BUA	British Undertakers' Association
LAFD	London Association of Funeral Directors
NAC(C)S	National Association of Cemetery (and Crematorium) Superintendents
NAFD	National Association of Funeral Directors
NCDD	National Council for the Disposition of the Dead

The Undertaker at Work: *Development and Service*

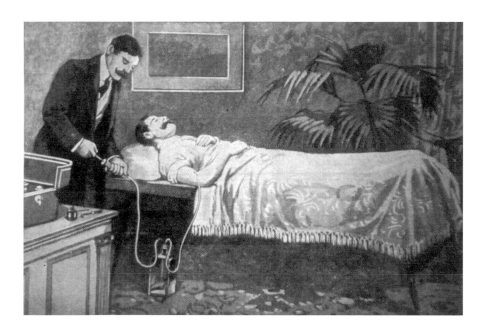

CHAPTER 1
THE PIONEERS OF PRESERVATION: THE INTRODUCTION OF EMBALMING

Although practised in Britain during the eighteenth century, embalming was a task outside the remit of the undertaker. However, to trace the starting point of modern arterial embalming, it is necessary to look to America in the late nineteenth century, where two events increased awareness of this method of preservation. The first was the embalming of President Lincoln following his assassination in April 1865. He was embalmed and transported on a lengthy journey from Washington DC to Springfield in Ohio for burial.[1] The success of the preservation indicated the value of embalming.

The second landmark was the American Civil War where Dr Thomas Holmes, proclaimed as 'The Father of Modern Embalming in the United States', embalmed the bodies of the battle-dead in preparation for transportation, invariably by rail, back to their hometowns.[2] He capitalised upon techniques used to preserve anatomical specimens as developed by the French anatomist, Jean Gannal, and also the Hunter brothers. At the end of the war, the efficacy of this preserving method was adopted commercially. Embalming schools opened and in classes lasting three or four days, lecturers demonstrated to undertakers simple injection techniques. One practitioner who took a particular interest in the subject was Dr Auguste Renouard, who went on to establish a number of embalming schools.[3]

This side of the Atlantic, the physicians Dr Benjamin Ward Richardson and Professor John Struthers were experimenting with cadaver preservation techniques. However, this was for the preservation of anatomical specimens and for bodies in coroner's mortuaries, rather than for commercial reasons.[4]

With the vast majority of people dying at home and also being kept there until the time of the funeral, the only non-invasive alternative for preservation was the ice chest. After the laying-out woman had concluded her rituals, the undertaker would take a measurement of the body and return with the finished coffin or an inner shell to then encoffin the body. If the body deteriorated, the coffin would simply be sealed.

In the last quarter of the nineteenth century, there is evidence to indicate the availability of embalming on a commercial basis. One of the first practitioners to advertise his service was Peter Itzstein, who was listed in the 1874 edition of Kelly's Directory for London.

▲ A body being embalmed
during the American Civil War

◄ Halford Mills was certainly not
adverse to self-promotion. This
image of an embalmed body after
six months treatment appeared
in 1897 on the cover of *TUJ* (*TUJ*)

EMBALMING

AND

PRESERVATION OF THE DEAD.

Photographed Six Months after Embalmed.

TO UNDERTAKERS.

MR. HALFORD L. MILLS,

Graduate of the United States School of Embalming,

Holds a Certificate of Qualification to Embalm.

Terms to the Trade, and any details that are desired, will be sent
on application to

31, Cambridge Place, Paddington, London, W.

Telegrams: "HALFORD MILLS, LONDON,"

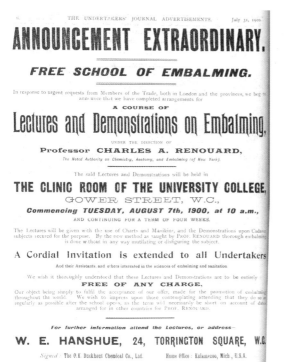

▲ 'An Extraordinary Announcement.' Professor Renouard's visit to England in 1900 (*TUJ*)

▲ The short-lived periodical *The British Embalmer* advertised details of correspondence and private tuition offered by Professor Felix A Sullivan

His one-line entry read: 'Bodies embalmed and petrified – will keep indefinitely. Terms on application.' From then onwards other embalmers were regularly listed, including Professor C H Garstin of Baker Street,[5] the funeral suppliers Dottridge Bros, and Halford Mills of the Reform Funeral Co. Mills was a prominent London undertaker, a pioneer of funeral reform and is reputed to be the first American-qualified embalmer offering his services in England.[6]

The turning point for the funeral industry was in 1900 when Dottridge Bros and the British Institute of Undertakers (see Chapter 2) invited Dr Auguste Renouard to London to give undertakers basic instruction in preservation techniques.[7] Classes commenced in August at University College London with some 25 people attending. A second class was organised and in total 30 graduated: being awarded a certificate of attendance, but not passing a practical or theoretical examination.

The Canadian, Professor Felix A Sullivan, arrived to conduct lecture tours throughout the British Isles.[8] As Chapter 2 notes, this development led to the founding in November 1900 of the British Embalmers' Society.

But to what extent was embalming practised? The London section of the BES reported its membership carrying out a total of 91 embalmments from when Professor Renouard commenced

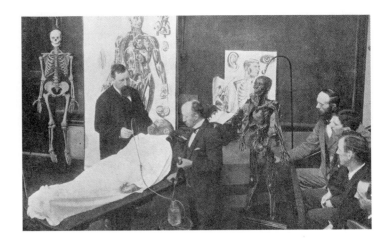

teaching to the end of 1902. Walter Uden from south-east London embalmed 45 cases after qualifying under Professor Sullivan. The records of J H Kenyon record six embalmments in 1900. Although the figures rose to 32 in 1901 and 56 in 1902, in the following decade a steady decline can be noticed.[9] Despite this discouraging situation, as Chapter 3 indicates, two members of the Kenyon family and other embalmers made a significant contribution following the Salisbury rail accident of 1906.

After the departure of Professors Renouard and Sullivan, the promotion of embalming was in the hands of British practitioners. In March 1906, Walter Uden led a three-week class in Preston under the auspices of the Anglo-American School of Embalming, while in 1907 the British Embalmers' Society ran a course of instruction.[10]

▲ Professor Sullivan conducting a demonstration, as seen in his book *The Champion Textbook of Embalming* published in 1897

▼ A photograph taken on the steps of University College London of the successful graduates of Professor Renouard's embalming class (*TUJ*)

A GROUP OF SOME OF THE STUDENTS OF
THE ANGLO-AMERICAN SCHOOL OF EMBALMING.
Lecturer and Demonstrator: PROFESSOR F. A. SULLIVAN.

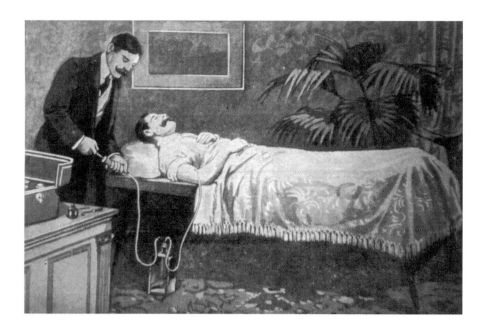

▲ In the early years, embalming was normally carried out at the place of death, usually the home. Dottridge included this image of a gloveless, moustachioed embalmer in their promotional literature

▼ Dottridge Bros also had a preparation room at their City Road premises, as shown in this image from 1903. Such facilities were the exception (*TUJ*)

The last US pioneer to visit Britain was in January 1910, when George B Dodge embarked upon a lecture/demonstration tour under the auspices of the Massachusetts College of Embalming. He was the author of *Sanitary Treatment of the Dead* and, like Renouard, Sullivan, A Johnson Dodge (George's brother) and other American practitioners, regularly contributed to *TUJ*.

After George Dodge's departure, a small number of pioneers were active in the years before and immediately after WWI. One pioneer was Arthur Dyer, who had attended Professor Dodge's class in London before receiving further instruction in the United States.[11]

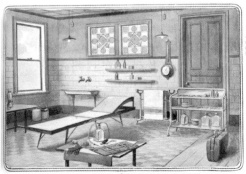

MODERN SANITATION.—Permanent and Temporary Embalming.

IN view of the increasing favour with which the process of Preservation of the Dead is being received in this country, **we are now prepared, on receipt of a telegram, to immediately dispatch a fully qualified and experienced Operator to any part of the Kingdom.** Testimonials may be seen, which have been received from various parts of the world, testifying to the efficiency of our work.

Terms upon application.

For the accommodation and use of the Trade, we have opened a large, well-ventilated Operating Room, fitted with the most hygienic appliances. The room has good light—an important consideration—and possesses every convenience for post mortem and embalming cases. In addition, we have a Mortuary, which can be arranged to meet the requirements of the Trade in cases of urgent removals from Hotels, Nursing Homes, Public Institutions, and transportation of bodies.

· LABORATORY & OPERATING ROOM ·

Students in this branch of the profession are carefully trained, and made practically conversant with all that relates to Arterial Embalming.

A Complete Price List of Laboratory Equipments, invaluable to all engaged in the Preservation of the Dead, can be had upon application.

SUNDRIES.—Portable Embalming and Post-Mortem Tables, Aprons, Pedestals, Carpets, Casket Covers, Antiseptic Transportation Sheets, Injection Bottles, Fumigators, Generators, Embalming Instruments and Outfits, &c., &c.

DOTTRIDGE BROTHERS, Ltd., Dorset Works, East Road, London, N.

▼ The announcement of George B Dodge's embalming classes to be held in London in 1910 *(TUJ)*

▲ Walter Uden's embalming class at Preston in 1906. Note the body on the table and also the anatomical wall charts *(TUJ)*

Massachusetts College of Embalming

(Chartered by the State as an Educational Institution),

GEO. B. DODGE.
President, Lecturer, and Demonstrator.

BOSTON, MASS., U.S.A.

London Headquarters:

Strand Palace Hotel, W.C.

ANNOUNCEMENT.

E. M. BARTLETT.
Vice-President, Lecturer, and Demonstrator.

PROF. DODGE will remain in this Country during **February, March,** and **April,** and will engage to instruct in any city in the United Kingdom **where those who desire instruction will form a Class.**

Terms will be given on application for a **Regular Term** of **Two Weeks,** or a **Short Term** of **Four Days.**

The best and most successful methods for **Temporary Preservation** and **Disinfection,** and the **Alco Process (without incisions)** will be taught at Short Sessions, and full instructions in Anatomy, Physiology, Sanitary Science and Embalming, and Disinfection and Disinfectants, at the Longer Sessions.

Private Students will be accepted at London.

Address for Particulars—

GEO. B. DODGE, Strand Palace Hotel, London, W.C.

London College of Embalming.

Under the supervision of W. O. NODES and J. HEATH.

We have arranged for our . . .

FIRST 1917 CLASS

TO TAKE PLACE IN

SHEFFIELD

On MONDAY, JAN. 8th, for ONE WEEK.

Our Slogan : "The Practical School."

IMPORTANT.—A week's scientifically arranged concentration on the subject will give you a marvellous grasp of the science, and will equip you with knowledge that will be useful to you in your life's work, and make Embalming easy. We do not pretend, however, you can know it all in this time, but you will be prepared when you go out to your cases, and your mind will be trained to easily acquire further knowledge, which only comes of experience.

The Lectures on the subject are given in language easily understood by all. The Only School that is able to guarantee plenty of Practical Work.

☞ JOIN OUR FIRST 1917 CLASS.

▲ An advert from January 1917 for the London College of Embalming *(TUJ)*

As a leading member of the British Embalmers' Society, Arthur Dyer joined with W Oliver Nodes in October 1913 to establish the London College of Embalming. Despite Dyer being called up for war service the following year, the college continued but with W Oliver Nodes teaming up with Joseph Heath to organise the Sheffield class of the London College in January, and again in September 1917.

Others active during this period include Morgan R Morgan of Neath in Wales and Albert Cottridge and James Goulborn from London.[12] Cottridge trained as a teacher before joining the family funeral business. He and Morgan were both heavily involved with the BES (and later the BUA) and also authors of textbooks and numerous articles on embalming. Enthused by his visit to funeral homes and embalming colleges in the US, Morgan R Morgan followed in the footsteps of Professor Felix Sullivan by embarking upon a lecture tour around Britain during 1923 and 1925.[13]

▼ L to R: Arthur Dyer in 1913. Two embalming pioneers, Albert Cottridge and Morgan R Morgan *(TUJ)*

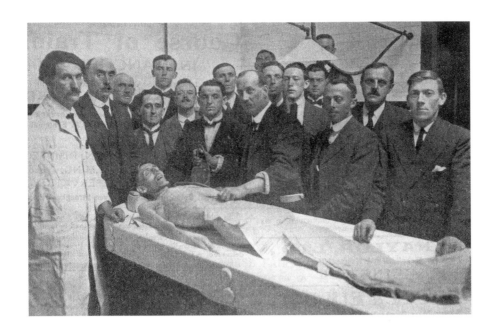

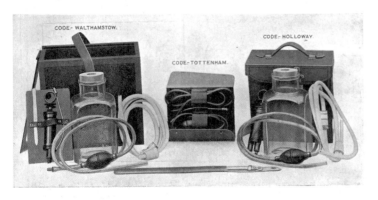

▲ An embalming demonstration conducted at the police mortuary at Preston as part of the BUA conference in 1921 *(TUJ)*

◄ Embalming equipment advertised in 1922 by Ingall, Parsons, Clive & Co. The embalming kits are all named after north London locations *(TUJ)*

▶ Ipsol embalming fluid was marketed by Ingall, Parsons and Clive in the 1920s *(BUA Monthly)*

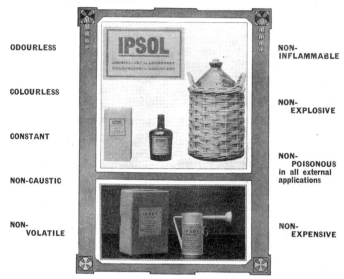

B.U.A. Monthly   XII   ADVERTISEMENTS

IPSOL

Disinfectant and Deodorant.

DESTROYS | FOUL SMELLS
HARMFUL GASES
DISEASE GERMS

Funeral Directors will find **IPSOL** an exceedingly useful article.

IPSOL absorbs readily all gases and prevents decomposition of any part of the body submitted to its contact.

IPSOL is invaluable to combat cadaveric emananations, to deodorize and purify soiled linen and asepticize all defections.

ODOURLESS

COLOURLESS

CONSTANT

NON-CAUSTIC

NON-VOLATILE

NON-INFLAMMABLE

NON-EXPLOSIVE

NON-POISONOUS
in all external applications

NON-EXPENSIVE

Descriptive Pamphlet and Prices from Sole Suppliers:

INGALL, PARSONS, CLIVE & Co., Ltd.,
206, Bradford Street, BIRMINGHAM.
And at LONDON, MANCHESTER, LIVERPOOL, BRISTOL, GLASGOW, &c.

Everything for the Undertaker.

▲ George Lear photographed in the 1950s

Towards the end of WWI, the availability of embalming tuition proliferated. The BUA's Manchester centre organised an 'Embalming Summer School' in June 1918, while in September, the Northern College of Embalming opened under the auspices of Messrs Godson, Scales, Ball and Morrison. Arthur Dyer and W Oliver Nodes also organised the first post-war 'Victory' class in Sheffield for ex-servicemen. During the BUA conference in 1918, an embalming demonstration was given at Hull City mortuary.[14]

As Chapter 2 discusses, the founding of the BIE in 1927 presented the opportunity for one organisation to focus exclusively on the promotion of embalming. However, instead of undertakers receiving training, they could utilise the services of a trade embalmer.

One such practitioner was George Lear. A native of Rochdale, he moved to London in the 1930s to set up a freelance embalming service.[15] He also founded a school and supplied equipment and fluids. Tireless in the promotion of the craft, he died in May 1954.

Using a freelance service meant that undertakers did not have to become a qualified embalmer or employ a full-time member of staff. At this time, many cases were embalmed at home or in public mortuaries as most undertakers did not have a fully equipped

▼ Another method of preservation was dry ice: frozen blocks of carbon dioxide. In the mid-1930s, Dottridge Bros marketed 'Drikold', while another brand was 'Cardice'. Available in blocks, it had a temperature of 110F below zero and could be used to freeze the abdomen and other parts of the body. However, it evaporated, and the blocks had to be replaced within twenty-four hours (*BUA Monthly*)

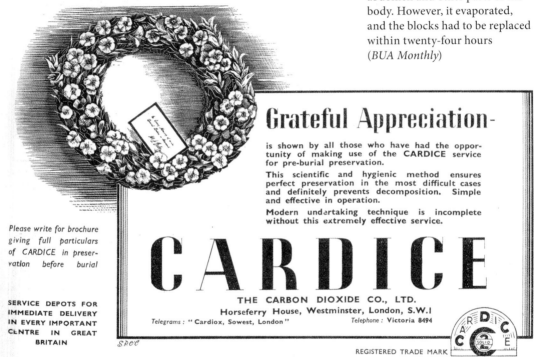

Please write for brochure giving full particulars of CARDICE in preservation before burial

Grateful Appreciation-

is shown by all those who have had the opportunity of making use of the **CARDICE** service for pre-burial preservation.

This scientific and hygienic method ensures perfect preservation in the most difficult cases and definitely prevents decomposition. Simple and effective in operation.

Modern undertaking technique is incomplete without this extremely effective service.

CARDICE

THE CARBON DIOXIDE CO., LTD.
Horseferry House, Westminster, London, S.W.1

Telegrams : " Cardiox, Sowest, London "   Telephone : Victoria 8494

SERVICE DEPOTS FOR IMMEDIATE DELIVERY IN EVERY IMPORTANT CENTRE IN GREAT BRITAIN

REGISTERED TRADE MARK

mortuary; certainly none had refrigerated accommodation until the 1950s.

Although the number of embalmments taking place in the first half of the twentieth century was very modest, a few notable examples can be identified, and five are outlined in the final part of this chapter.

Sir Arthur Sullivan is best known for his collaboration with the librettist W S Gilbert to produce such works as *HMS Pinafore*, *The Pirates of Penzance* and *The Mikado*. In all, 23 operas form part of Sullivan's output, which also included 13 major orchestral works, choral works and oratorios, two ballets, numerous hymn tunes, parlour ballads, carols and piano and chamber music.

Sullivan suffered from recurrent kidney disease and, from the 1880s, this necessitated him to conduct whilst sitting down.

▲ Sir Arthur Sullivan (1842–1900)

He died following an attack of bronchitis at his home, 1 Queen's Mansions, Victoria Street, London, on 22 November 1900. The cause of death was 'bronchitis, 21 days; cardiac failure'.[16]

Sir Arthur expressed a wish to be buried beside his parents in Brompton Cemetery. He detailed his funeral instructions (dated 18 August 1882) in a sealed envelope, with the first three being:

- *I wish my body to be embalmed before burial. Let nothing prevent this being done.*

- *My funeral is to be conducted in the same manner as that of my dear Mother, and if possible, by the same undertakers.*

- *My body to be buried in the same grave with my Father, Mother and Brother in Brompton Cemetery.*[17]

Quite why Sir Arthur Sullivan requested embalming is unclear. At the time that he gave these instructions – eighteen years prior to his death – few embalmments would have been carried out. It could well have been stipulated for fear of premature burial; a concern that lasted well into the twentieth century.[18]

Sir Arthur Sullivan was embalmed by Alfred Speight, an undertaker of Manningham Lane, Bradford.[19] Speight had attended the first class of embalming lectures and demonstrations by Professor Felix Sullivan in October 1900. Although resident in Bradford, it is likely that he happened to be in London at the time of Sullivan's death and so came to embalm the composer. The body was kept at home until the funeral, which took place on 27 November 1900. The composer had stated that he wished to be buried at Brompton Cemetery and arrangements progressed to the extent that the family grave was reopened. The Dean and Chapter of St Paul's Cathedral then consented to an interment preceded by a service in the Chapel Royal, at the permission of Queen Victoria, a gesture that was accepted by Sullivan's family. The coffin was lowered into a vault directly below the choir. Speight's name continued to be associated with embalming and the activities of the fledgling British Embalmers' Society for some years after 1900.

Sir Henry Irving was renowned as the finest Shakespearian actor of the nineteenth century and was the first actor to be awarded a knighthood. He died on Friday 13 October 1905 – less than two hours after appearing as Becket on stage at the

Theatre Royal in Bradford – when he collapsed in the foyer of the Midland Hotel. He died in the arms of his devoted dresser, Walter Collinson. Sir Henry was aged 67 years. He was embalmed by Joseph Jarrom, and *TUJ* provides some information on this case:

▲ Sir Henry Irving (1838–1905)

> *'At the first meeting of the newly-formed British Embalmers' Society, Incorporated, held in Manchester on the 17th of October, Mr JW Bellarby, who presided, welcomed Mr Joseph Jarrom of Bradford, who, he said, had had the honour which they also considered an honour to the Society, of embalming the remains of the late Sir Henry Irving.*
>
> *Mr Jarrom, who represented Messrs Holmes & Co, and is quite a young man, said the case presented nothing extraordinary, it was one of the simplest, but there were two or three useful features which cropped up. He was not called in until about nine o'clock in the morning, and the remains left for London shortly before ten o'clock the same night.*
>
> *The relatives and friends who saw the remains, including Mr and Mrs George Alexander, Sir Squire Bancroft, Mr HB Irving, Mrs Laurence Irving, and others, were all pleased with the result of the embalming, and marvelled at the perfect naturalness of the features, which had no more semblance of death than though he has just dropped asleep.*
>
> *The embalming fluid used was the celebrated OK Fluid supplied by the Buckhout and Breed Chemical Co.'[20]*

Sir Henry's body was transported from the Midland Hotel to the Great Northern railway station in Bradford.[21] Again, *TUJ* furnished more information about the arrangements in London:

> *'The remains were received in London by Messrs Mills & Sons Ltd, undertakers, and conveyed to Sir Henry's late residence in Stratton Street, Piccadilly, whence on 17th they were conveyed to their private mortuary in Cambridge Place, Paddington. On the morning of the 18th, the coffin (A "Windsor", from the Manchester works of John Yates & Sons) was reopened to permit a sculptor and his assistants to take a wax impression of the dead actor's features. A few hours later a simple-looking hearse drew up at the mortuary. Into it six*

men bore the body, which was contained in a plain, shallow shell, and thickly enfolded in a blue pall. The receptacle bore little resemblance to what it actually was, and the hurrying passers-by gave it no heed. Then the six men entered a black-visaged van, the closed, windlowless side of which hid them from view. Hearse and van then set out for Golder's Green Crematorium, Finchley Road. Few noticed the little procession on its long journey, and no one accompanied it except the undertakers, their assistants, and a reporter...

No sign of recognition indeed was forthcoming until the gates of the crematorium were reached. A solitary carriage stood outside the chapel door. From it there had just alighted Mr Laurence Irving and "Walter", the latter for thirty years Sir Henry's valet. The shell containing the coffin was deposited on a table-like structure, at one end of which was a pair of iron gates, flanked with beautifully-tinted marble columns. Above was the inscription, "Mors Janua Vitæ." The two mourners stood alone. Presently from a side door there emerged a uniformed attendant. For a moment he awaited the signal. Then his hand touched a small steel lever, and the coffin very slowly passed into the incinerating chamber.

An hour and half later the few ashes – all that remained of the admired tragedian, who was known to and beloved by countless people – were gathered into an urn, which was reverently taken to the house of the Baroness Burdett-Coutts in Piccadilly.'[22]

▼ L to R: General William Booth (1829–1912). Mr S Diamond of Dottridge Bros who embalmed General Booth

The urn was placed in a full-sized coffin where it lay in state. Sir Henry was among the first to be cremated prior to burial in Westminster Abbey. His ashes were buried in the chapel of St Faith on Friday 20 October 1905.

Founder of The Salvation Army, General William Booth died on 20 August 1912 at Hadley Wood in Hertfordshire. He was embalmed by Mr S Diamond of Dottridge Bros, who also managed the complex funeral arrangements. The general's body was transported to Clapton where it rested in the Congress Hall; two days later the coffin was taken to Olympia in west London, where a funeral service was held on 27 August, attended by an estimated 25,000 mourners. The coffin was then transferred to the Army's headquarters in Queen Victoria Street, and finally to Abney Park Cemetery in Stoke Newington, where General Booth was buried with his wife, Catherine.

In Congress Hall, the general lay in a heavy oak coffin that rested on a marble catafalque. The foot of the coffin was covered with Army colours, while a glass panel in the upper part of the lid revealed General Booth's face.

Mr S Diamond commenced working with Dottridge Bros in 1894. He studied embalming under both Professors Renouard and Sullivan and also George B Dodge. He was an early member

▼ Madame Adelina Patti (1843–1919)

of the BES and was also a founder member of the BIE in 1927; he was made an honorary member of the Institute upon retiring from Dottridge Bros in December 1945.[23]

Madame Adelina Patti was a famous opera singer who sang at the world's leading opera houses in the 1880s and 1890s. Although she was born in Spain and her parents were Italian, she married in Wales and lived for much of her life at Craig-y-nos Castle near Ystradgynlais, Brecon. She earned £1,000 for a single performance at the height of her fame, which was a huge amount of money for the time – Madame Patti was very generous and did a great deal to help local people in Wales.

After losing consciousness, she died at 9.30 a.m. on 27 September 1919.

In an obituary entitled, 'Death of Patti: The "Bel Canto" School' published in *The Times* two days after her death, it was stated that, 'The funeral is to take place in Paris, but due to the strike the arrangements are uncertain.'[24] Transport to France with a delay due to a strike necessitated embalming. This was carried out by Morgan R Morgan, who provided a lengthy account of this long-term embalming in *TUJ*:

> *'Madame Patti was a lady over 80 years of age [she was 76] and in a remarkable state of preservation for one of her years. I was called shortly after death and had to travel 25 miles to Craig-y-Nos Castle, a beautiful country house in the heart of the Welsh Mountains, arriving there within 24 hours of death. I then proceeded forthwith with the care of the body in which I was assisted by two hospital nurses and in the presence of certain ladies-in-waiting. My first treatment was that of a spray of Necrosan for the whole body, followed by a massage treatment of the head and neck with Necrosan cream. I then proceeded with embalming, using the axillary artery and vein by what is known as the arterio-venous system, by means of which I secured a thorough circulation. Upon one occasion I observed a blood-clot on the chin, which was readily removed by means of massaging the chin in the direction of the jugular vein and withdrawing by means of the venous tube. The whole time taken up in the treatment was two hours. The cosmetic effect of the embalming upon the face gave to the body a rejuvenating appearance, removing the wrinkles of*

age....The body remained on view in the Castle for three or four days before it was enclosed in the three coffins, according to custom.

Did you go back to the body during those three days? Did you hear how the remains kept?

Satisfactory reports were received regarding the condition and, therefore, there was no necessity for my paying a subsequent visit. In about three weeks the remains were removed from the Castle to a crypt in St Mary's Church, Kensal Green, London, and remained there until May 28 [29], 1920, when the body was shipped to Père Lachaise Cemetery, Paris, where the final interment took place, nine months after death. Satisfaction was expressed by all those concerned with the manner in which the arrangements had been carried out.

How did those who were present regard your work, they having not been used to embalming?

One of the nurses gave expression when she saw the nature of the instruments which I carried with me, remarking, "Are you going to perform an operation?" taking me evidently in the light of a physician.

What did you say to her?

In some respects, yes, but in the event no organs would be removed or any injuries done to the body apart from opening as to locate the necessary arteries and veins, so that I could perform the preserving function which I was called upon to do.'[25]

Madame Patti's body was brought to Paddington station in a special saloon on Friday 24 October, 1919 and then deposited in the crypt under the chapel in St Mary's Roman Catholic Cemetery at Kensal Green. The coffin remained there until 29 May 1920 before transportation to Paris for interment in Père Lachaise Cemetery.[26]

King George V died on 20 Jan 1936 and an account in *TUFDJ* details the embalming arrangements:

'The coffin of Sandringham oak, bearing the inscription "George Frederick Ernest Albert Windsor, 1865–1936" was removed from Sandringham House to the Village Church on Tuesday evening. Already the body has been embalmed by Mr LV Weaving of Messrs Garstin and Sons of Wigmore Street, W,

▲ W Oliver Nodes: First
president of the British Institute
of Embalmers, he assisted with
the embalming of George V

◄ George V photographed
in 1910. He was embalmed
at Sandringham

*with the co-operation of W Oliver Nodes, a Past
President of the British Institute of Embalmers. This
Royal act in recognising the propriety of the science of
embalming – hitherto performed to rulers of the land
by the King's physicians and doctors – is one that is
deeply appreciated by the profession as a whole, being
regarded as a gracious tribute to the years of efforts of
those who have striven for the more general adoption
of the science for the sanitary care of the dead.*[27]

The records of William Garstin funeral directors confirm this
information. Their registers reveal an entry for 'His Late Majesty
King George V':

*Principal and 2 embalmers attending at Sandringham
to embalm his late Majesty King George V.
Motor car for conveyance to Sandringham
Complete £75*

The account was settled by the Secretary of the Privy Purse at
Buckingham Palace on 11 March 1936.[28]

1. See EC Johnson, GR Johnson, MJ Williams (1990) 'The Origin and History of Embalming and History of Modern Restorative Art' in Mayer RG *Embalming. History, Theory and Practice* Stamford: Appleton & Lange. See also Johnson EC (1975) 'Cattell's Skill with Lincoln's Remains Publicized Embalming' *Casket and Sunnyside* September Vol 105 No 9 pp16,18, 20, 53–54. Huntington R & Metcalf P (1979) *Celebrations of Death: The Anthropology of Mortuary Ritual* Cambridge: Cambridge University Press pp206. See also Craughwell TJ (2007) *Stealing President's Lincoln Body* Cambridge MA: Belhnap/Harvard

2. For a full account of the development of embalming during the American Civil War see 'Civil War Embalming' Johnson EC (1983) *The Embalmer* Vol 26 No 4 July p25. See also 'The Embalmer Jean Nicolas Gannal (1791–1852)' *TNFD* 1937 March pp227–228

3. See *The Embalmer* (1987) Vol 30 No 5 October pp5–6 for a full account of Auguste Renouard's life (1839–1912)

4. Parsons B (2005) 'Benjamin Ward Richardson: A Forgotten Pioneer of Embalming' *The Embalmer* Vol 48 No 4 pp16–20 and Struthers J (1890) 'On the Preservation of Bodies for Dissection' *Edinburgh Medical Journal* Vol 36 September pp297–303. See also 'Sir John Struthers on Embalming the Dead' (1898) *TUJ* July p112. For previous centuries see Dobson J (1953) 'Some Eighteenth-Century Experiments in Embalming' *Journal of the History of Medicine and Allied Sciences* Vol 7 pp431–441

5. 'A Famous London Firm: Messrs W Garstin & Sons Ltd' (1936) *TUJ* December pp393–395. William Garstin & Co was responsible for embalming Lady Dilke. See Parsons B (2005) *Committed to the Cleansing Flame: The Development of Cremation in Nineteenth-Century England* Reading: Spire Books

6. Parsons B (2006) 'Halford Lupton Mills' *The Embalmer* Vol 48 No3 pp15–18

7. Parsons B (1995) 'The Pioneers of Preservation: The Development of Embalming in the UK' *The Embalmer* Vol 38 No 1 pp16–24

8. See Johnson EC (1974) 'Life and Times of Felix Sullivan, Noted Embalmer' *Casket and Sunnyside* Vol 104 No 7 pp18, 20–21, 23 and 49, and 'The Lives of the Teachers: Pioneer Embalming Educators of America. Prof Felix A Sullivan' (1967) *The De-Ce-Co Magazine* February Vol 59 No 1 pp14 and 22, 'The Late Professor FA Sullivan' (1930) *TUJ* August p262

9. Parsons B (2005) *JH Kenyon: The First 125 Years* Worthing: FSJ Communications p42

10. Parsons B (2006) 'Portrait of an Embalming Pioneer: Walter Uden' *The Embalmer* 2006 Vol 49 No 1 pp20–22. See also Uden W (1910) 'Embalming – A Retrospective' *TUJ* June pp137–138

11. Dyer AB (1949) 'On Ideals' *FSJ* January pp21–22

12. 'Mr James Goulborn' (1943) *TNFD* May p416

13. Morgan MR (1919) 'American Impressions' *TUJ* November pp336–337 and December p370 and (1920) January pp25–26. See also (1923) January p9, March p120 and May p167 and (1923) October p346 and November p407 and (1924) October p343. See also 'MR Morgan's Lecture Tour' (1925) *TUJ* March p87

14. 'The Demonstration and Lecture' (1918) *TUJ* July p177

15. School of Embalming' (1933) *TUJ* September p299. See also Parsons B (2004) 'George Lear: A fiftieth anniversary tribute' *The Embalmer* Vol 47 No 1 pp11–15

16. Young PM (1971) *Sir Arthur Sullivan* London: JM Dent & Sons p259. See also *The Times* 23 November 1900

17. Sullivan H & Flowers N (1927) *Sir Arthur Sullivan: His Life, Letters and Diaries* New York: GH Doran pp 328–329. See also A Jacobs (1984) *Arthur Sullivan: A Victorian Musician* Oxford: Oxford University Press p398

18. See J Bondeson (2001) *Buried Alive: The Terrifying History of Our Most Primal Fear* New York: WW Norton & Co

19. See *TUJ* (1900) November p157

20. 'A Bradford Member of the BES and the Late Sir Henry Irving' (1905) *TUJ* November p264. Jarrom died on 23 March 1959 aged 89. Messrs J Holmes closed in 1921 and Jarrom left embalming to enter another career

21. Irving L (1951) *Henry Irving: The Actor and his World* London: Faber & Faber pp 671–673

22. 'The Passing of Sir Henry Irving' (1905) *TUJ* November p276. See also Richards J (2005) *Sir Henry Irving: A Victorian Actor and his World* London: Hambledon & London pp1–3

23. 'Retirement of Mr S Diamond. Fifty years as Funeral Manager to Messrs Dottridge Bros Ltd' (1946) *FSJ* March pp92–93. See also 'Honoured Names' (1948) *FSJ* July p322 and 'Funeral of General Booth. A Great Pageant in Memory of a Great Man' (1912) *TUJ* September pp257–260. See also 'Sanitization of the Dead' (1922) *TUJ* February pp55–56

24. *The Times* 29 September 1919

25. 'Embalming of Madame Patti' (1920) *TUJ* October p312

26. 'Personal and General' (1920) *TUJ* June p184. See also Cone JF (1994) *Adelina Patti: Queen of Hearts* Aldershot: Scholar Press p264

27. 'His Late Majesty King George V' (1936) *TUFDJ* February pp39–40

28. Garstin's records are held at Westminster Archives Centre. For a short profile of the firm see Parsons B (2005) *Committed to the Cleansing Flame: The Development of Cremation in Nineteenth-Century England* Reading: Spire Books pp115–116.

CHAPTER 2
INSTITUTE, SOCIETY AND ASSOCIATION

Organisations representing the undertaker have been an important part of the landscape of funeral service between the end of the nineteenth century and 1950, and a number emerged to represent different sectors and interests. Whilst embalming was identified as the route towards elevating the status of the occupation, the formation of the British Embalmers' Society in 1900 provided the springboard for other organisations including the British Undertakers' Association, which ultimately became the National Association of Funeral Directors. However, the promotion of sanitary treatment of the dead became overshadowed by issues between undertakers and carriagemasters before the focus shifted to statutory registration.

THE BRITISH INSTITUTE OF UNDERTAKERS

Organisations representing the interest of those engaged in the disposal of the dead can trace their roots back to the eighteenth century, when the formation of the Upholders' Company was closely followed by the United Company of Undertakers.[1] Despite the expansion of the sector in the nineteenth century, no representative association appears to have been in existence until 1894, when a group of London undertakers rallied to defend colleagues against a libel action brought about by a coroner's officer.[2] By October 1898, this had become the British Institute of Undertakers (BIU). The membership was largely London-centric; some such as James Hurry, Henry Sherry, Henry Kellaway and Walter Uden would have a long association with the different organisations that emerged. Henry Sherry believed in the need for an association that '...should watch over the interests of the trade, and which ... would earn the respect of municipal and governing bodies.'[3] The issue of registration of undertakers was discussed at one meeting with James Hurry proposing that, 'The BIU do all in its power to petition or otherwise to get Parliament to make some form of compulsory registration.'[4] However, despite claiming to be '... recognised as the central authority of the undertaking profession ...' the Institute achieved little, although dialogue had been commenced with colleagues in Leeds and Liverpool. Nevertheless, Henry Sherry believed the BIU had a future and looked to the rise of similar organisations in America as the model for gaining recognition. At a meeting in November 1899,

▼ The British Institute of Undertakers logo. Its crest was the horse rampant

he urged members to '... follow their cousins across the water ... organize, educate, legislate.'[5] He also believed that professional development would come through advocating sanitary practices.

By the beginning of 1900, however, it was clear that the Institute was foundering, as *TUJ* observed:

> *'It is a lamentable fact that those who have been endeavouring to make non-MBIUs believe that a mighty power was rising up among them should have been confronted by such humiliating failure. We had thought that long ere this, according to its own gospel, that the BIU would have reconciled the trade unto itself. But it had not proved to be, and even those of its rank and file seem to be either lagging or falling out.'[6]*

Skirmishes between the editor of *TUJ* and Henry Sherry about criticisms of inactivity detracted from identifying what the Institute was achieving. By December 1900, a restructuring had been proposed but was shelved, as had been a congress scheduled for the following year.

As indicated in Chapter 1, Professor Renouard's visit to run a course of embalming lectures caught the interest of a number of undertakers. Attended by representatives from many London firms, the 'phenomenal success' of the course led to a further

▾ Henry Sherry (1850–1933). London undertaker, and first president of both the British Embalmers' Society and the British Undertakers' Association *(TUJ)*

MR. H. A. SHERRY.

class in December presided over by Professor Felix A Sullivan. Through recognition of its sanitary benefits and by giving undertakers technical skills, embalming provided a strong basis for collaboration and, with Professor Renouard's encouragement, the opportunity to unite was fostered. *TUJ* enthused about the idea for an embalming organisation and the creation of an association was proposed in a leader.[7] In November 1900, a new organisation was founded with Henry Sherry as the chairman.[8] At Halford Mills's suggestion, it was called the British Embalmers' Society.

THE BRITISH EMBALMERS' SOCIETY

With a primary objective to '... encourage the practice of modern embalming, sanitation and care of the dead in the British Isles', membership of the BES was through examination: the letters MBES could be appended behind a member's name.[9] With Professor Sullivan leading classes in towns and cities around Britain during 1900 and 1901, this was the most pro-active way of recruiting members. Midway through 1901 it was reported that the Society had unsuccessfully attempted to get a Board of Trade licence; there was a second attempt in 1903 but it was met by a similar outcome.[10] During 1903, national meetings were held but with Professor Sullivan back in Canada, it was up to members of the BES to promote embalming and organise training.

Against this background of activity, the BIU appeared to be dormant.[11] In March 1902, Sherry had described the Institute as having 'fallen asleep' and wished that the BES '... would extend its sphere of labour to a fostering of the funeral interest of the profession.'[12] Again, he restated the need for a cohesive voice:

> *'That there is to-day as great, if not a greater, need for unity in the undertaking trade, no one who reasonably reviews the existing conditions can deny. When one thinks of such matters as....so-called 'Funeral Reform';...'Premature Burial';...the coroner's officer;...that clerical adversary, the cemetery chaplain...the all too officious cemetery official, one is at once convinced of the need of unity...'*[13]

Furthermore, commission payments for burials in parochial cemeteries had been stopped and a coroner in south London had actively prevented bodies being kept on undertakers' premises as he believed them to be '... insanitary and unfit for reception of the dead ...'[14] The Government's Commission on Death Certification

Convention of Undertakers and ~ Funeral Trades Exhibition in London.

THE FUNERAL FURNISHERS' ASSOCIATION (London) have arranged
for a CONVENTION and EXHIBITION to be held on

WEDNESDAY, THURSDAY, & FRIDAY, June 7th, 8th, & 9th, 1905,

. . IN THE . .

Great Hall of the Northampton Institute,

ST. JOHN STREET ROAD, CLERKENWELL, LONDON

(Three minutes from " Angel," Islington, convenient and easy of access from all London Railway Termini).

The whole of the Stall Spaces have been applied for by Exhibitors, who will display

The Latest Developments in the Supply of Undertakers' Requirements.

The Exhibition will be opened by **JAMES T. SLATER, Esq.** (of Messrs. Bedford, Son, & Slater),
at 2 p.m., on Wednesday, June 7th, **W. KNOX, Esq.,** presiding.

The Convention, which should be attended by every Undertaker in the Kingdom, will be opened with prayer
by the REV. M. C. RICHARDSON, REV. GEO. CURTIS, and the REV. J. H. BENFIELD respectively.
(Church of England).   (Roman Catholic).   (Nonconformist).

DURING THE WEEK

The Funeral Furnishers' Association (London)

Will hold their FIRST ANNUAL GENERAL MEETING, and

The British Embalmers' Society

Will hold their FIFTH ANNUAL GENERAL or NATIONAL MEETING, beside their
respective Committee and National Council Meetings.

Wednesday, 7.30—CHAS. PORTER, Esq. (Liverpool), will Lecture on "How to Form Undertakers'
Associations."

Thursday, 7.30—HENRY A. SHERRY, Esq. (London), will give an Address on "Basis of Constitution
for the National Association."

Friday, 3—Dr. A. WYNTER BLYTH, M.R.C.S., F.C.S., F.I.C., will speak on "The Hygiene of the Death
Chamber."

The Funeral Furnishers' Association, in conjunction with the British Embalmers' Society, have arranged for
A BANQUET, followed by a MUSICAL ENTERTAINMENT AND DANCE, to be held at the Holborn
Restaurant, on FRIDAY, JUNE 9th, 1905. Reception 6.30 p.m. Banquet 7 p.m. prompt. Evening Dress optional.
Tickets 10/6, including light refreshments. As number of Tickets will be limited, early application should be made to

H. A. KELLAWAY, Hon. Sec., 250, Camberwell Road, S.E.

◄ The announcement of the Convention of Undertakers and Funeral Trades Exhibition in London, June 1905 *(TUJ)*

was underway, to which undertakers believed that they should have an input.

By March 1904, Henry Sherry acknowledged that the BIU had been defeated:

> 'That body has gone to sleep, but should wake up, if it is only for the purpose of picking up the scattered threads, and completing the making of that cord which was intended to bind together all the undertakers of this kingdom in one common organization.'[15]

THE BRITISH UNDERTAKERS' ASSOCIATION

Early in 1903, the Manchester-based North of England Funeral Undertakers' Association had been formed, and by January the following year, it had 40 forty members.[16] Its secretary revealed that the objectives of the association were to protect the interests of undertakers, rectify grievances and increase efficiency of members through ensuring they received '... some personal benefit ... by being enlightened on trade topics.'[17]

The success experienced in the North clearly encouraged their counterparts in London. A gathering presided over by James Hurry in March 1904 led to the formation of the Metropolitan Funeral Furnishers' Association whose objective was:

> 'To establish in the Metropolitan area an Association open to Master Undertakers and Funeral Carriage Proprietors, Monumental Masons and members of Allied Trades, for the purpose of improving the conditions of the trade; to watch Municipal, Hospital, Cemetery, local and similar Authorities in the interest of the Trade.'[18]

One of the motions advanced was the adoption of a minimum charge for funerals.

An important development was the announcement by the Manchester Association that an 'Undertakers' Convention and Funeral Trades Exhibition' would be held in September 1904.[19] Suppliers of coffins, furnishings, linings, memorial cards, trestles and artificial wreaths signed up to exhibit alongside carriage builders and suppliers of embalming fluids. Just prior to the convention, an association was formed in Liverpool.[20] Two of the papers presented at the Manchester exhibition were 'Embalming and Sanitation: Their value to an up-to-date undertaker' and 'The Modern Undertaker'.[21] However, it was Henry Sherry's paper entitled, 'The Need of a National Association of Undertakers' in which he analysed the notion of '... elevation of our status ...' through the adoption of embalming that made the greatest impression. He also advocated the need for legislation to license qualified undertakers.[22] The last contribution to the debate was a paper delivered to the Liverpool Association in November by Charles Porter, who spoke of 'The Advantages of Association.' Quoting from scripture, Alexander Pope, Shakespeare and Aurelius, Porter expressed a realistic objective:

> 'Rivalry there must be. Competition there must be; but where the Association will help is, not in restricting the rights of any individual member, but in adopting safeguards and preventatives of those things which are really and truly injurious to the best interests of the

business in general, and of each individual member in particular.'[23]

With much discussion about a London convention, *TUJ* commented:

> *'... one thing is certain; the undertakers of Great Britain never had such a pregnant chance before them as they have today. The ball is at their feet ... We know they want to raise the status of the trade and to rid it of many evils and anomalies ... We see the result in the associations that are springing up on every side. And the movement is spreading. During 1904 the progress has been enormous. Who shall say what it will be in the current year?'*[24]

The proposal of a London convention was discussed by the London Funeral Furnishers' Association in March 1905, and again suppliers expressed their support. Discussion on the formation of a National Association of Funeral Furnishers was urged, while Henry Sherry outlined the possible constitution of a National Union of Associated Undertakers. He stated that one of the first tasks for the organisation would be to watch the Bill proposed by The Society of Prevention of Premature Burial and to ensure that undertakers' interests were looked after.[25] By this time, there were eight undertakers' associations and four branches of the BES. In parallel with the developments, Sherry proposed winding up the BIU.

At the convention held on 7 June, Charles Porter read a paper on, 'How to Form a Funeral Furnishers' Association', in which he spoke of the role of officials, committees, the administration, objectives, rules, regulations and bye-laws. The following evening two papers were given. The first was, 'The Hygiene of the Death Chamber' by Dr Alexander Wynter Blyth, the medical officer for health for the Borough of Marylebone, while in the second Henry Sherry outlined his vision of 'A National Organization of Undertakers.'[26] He suggested a name, outlined objectives, proposed holding an annual convention, the publication of a yearbook, and the formation of a legislative or parliamentary sub-committee. The resolution of the convention was overwhelmingly positive and, at a meeting on 27 July in Birmingham, representatives from London, Liverpool, Preston and Manchester proposed a

▲ A laurel wreath frames the BUA's logo

▶ The BUA/BES Code of Ethics introduced in 1908 (*TUJ*)

CODE OF ETHICS

APPROVED BY THE

British Undertakers' Association and the British Embalmers' Society

(INCORPORATED).

1. As an undertaker on entering the business becomes thereby entitled to all its privileges, he incurs an obligation to exert his best abilities to maintain its honour and dignity, to extend its usefulness, and to exalt its standing.

2. Secrecy and delicacy, when required by peculiar circumstances, should be strictly observed. The obligation of secrecy extends beyond the period of our professional services.

3. None of the privacies of personal and domestic life should ever be divulged.

4. An undertaker should rely chiefly on his professional abilities and acquirements for the development of his business, and should discourage advertisements that tend to loudness and competition, looking forward to the time when such advertisements will be considered unprofessional.

5. When two undertakers are called at the same time to attend the same case, both should show a willingness to withdraw, leaving the choice with the family.

6. An undertaker should not shrink from the faithful discharge of his duties in case of epidemic and contagious diseases.

7. When an undertaker is called in case of sudden death or accident, because the family undertaker be not at hand, he should offer, if the family so desire, to resign the case to the latter, who should remunerate him for services rendered.

8. When an undertaker accompanies the remains and funeral party to a distant place, and if the remains are placed in care of another undertaker, anything he may do after that should be as a friend of the family, and as assisting the undertaker in charge of final arrangements.

9. When an undertaker orders from a distant place a corpse to be prepared and shipped to his care, all proper expenses should be charged to the undertaker giving the order, and it should be considered a professional obligation, and payment made at once.

10. Touting and soliciting for funeral orders is derogatory to the profession and should be rigorously discouraged.

constitution and bye-laws. The 'British Undertakers' Association' was adopted, with local areas becomes 'centres'.[27] Objectives were also formulated, with the first eight being:

- *To unite all existing local undertakers' Associations, and to promote the formation of other branches in the different parts of the British Isles*

- *To raise the tone and status of the calling and to promote the general education of its members in trade matters*

- *To watch over legislative measure which may affect or tend to affect, the best interests of members of the Association*

- *To advise associated branches in matters of local difficulty*

- *To work for and endeavour to obtain registration, so as to put the calling on the basis of a profession*

- *To organize conventions, at which propositions for the advancement of the interests of undertakers may be brought forward and discussed*

- *To aid all influences calculated to increase the usefulness and improve the recognition and the utility of undertaking as a profession, and to incorporate reforms wherever necessary*

- *The Association shall consist of all affiliated Associations, which shall be represented by delegates.*[28]

▲ J. R. Hurry – Stratford-based undertaker, third president of the BUA and national secretary of the association from 1920 to 1934 (*TUJ*)

Henry Sherry was elected first president of the British Undertakers' Association and James Hurry as the national secretary.[29]

THE BUA BECOMES A TRADE UNION

The immediate task of the BUA was to recruit members. Meetings of its regional 'centres' were arranged, as was an annual convention. Topics discussed in the early years included touting and secret commission, premature burial and embalming. Much debate centred around procedural and constitutional matters, and two important developments did occur. First, the introduction of a Code of Ethics in 1908, although it received little publicity and there was no scope for ensuring its adherence. Secondly, the BUA and BES amalgamated in 1916. Both organisations already had significant overlap of membership. [30]

During these formative years, membership was not large: between 1916 and 1917 there were only 203 members. As outlined in Chapter 4, it was during WWI that the association started to face an uncertain future due to the small membership compounded by the shortage of staff, the closure of businesses and restrictions on supplies impacting on the economic fortunes of the industry. The promotion of embalming was also curtailed and attendance at meetings was minimal. It was, however, the former factors that prompted the founding in 1916 of the London Funeral Carriage Proprietors' Association (LFCPA). Its objective

▶ W Oliver Nodes (sixth left) –
North London undertaker and
first president of the British
Institute of Embalmers

was to set a minimum price for the hiring of horse-drawn
carriages to undertakers.[31] Horace Kirtley Nodes urged that:

> '...all carriagemasters should become members. We
> are not out to form a ring, but on the contrary to make
> a strong combination as that there shall be no place for
> the undercutter, but every firm, of whatever size, shall
> have the opportunity of making a fair profit.[32]

By April the following year, the LFCPA had agreed a tariff of
hiring charges with the BUA.[33] Furthermore, carriagemasters
that were members of the LFCPA could only hire vehicles to
members of the BUA. This meant that unless they possessed their
own horses and hearses, all undertakers had to be member of the
BUA, otherwise they could not be supplied by LFCPA members.
It was this 'protection' of members that resulted in the Miller case
(see Chapter 6).

A further significant development was in December 1917
when the BUA became registered under the Trades Union Act
1871. This enabled the BUA to have greater control over their
members. As a summary noted:

> 'One of the powers given to them [the BUA] was to
> impose restrictive conditions upon members; they
> had powers also to improve the position and status of
> the trade, also to arrange for the higher education of
> their members.'[34]

This change in status for the BUA had an immediate effect on membership which increased from 700 by the time of the 1918 conference to 2,000 a year later; there were 660 members in London. Over the next few years, centres were formed in most urban areas: Bolton was one of the first outside London to negotiate a price agreement.[35]

Concurrent with this development was the formation in 1917 of the British Funeral Workers' Association (BFWA).[36] Founded by Thomas Kingston, its objective was protecting the interests of all funeral workers through improving working conditions, introducing a 5½ day week and enhanced wages.[37] The first wage negotiation between the union and the BUA took place in 1918, by which time it had a membership of 1,200.[38]

The impact of the BUA becoming registered as a trade union was that much time was taken resolving issues between carriagemasters, undertakers and their employees. Although within a few years the carriagemasters in London would unite in membership, it was the decline in sanitary education that would lead to the emergence of a further trade association.[39]

THE BRITISH INSTITUTE OF EMBALMERS

As chapter 1 noted, in the years following the end of WWI the BES was losing its way. Notwithstanding the curtailment of educational activities between 1914 and 1918, it would also appear that its function had been marginalised by its 'parent' organisation. As a trade union, the BUA was focused on protection of its members and consequently the BES had '... lost its individuality and also its charter of incorporation.'[40] In February 1926, the London Area Embalmers' Section of the BUA debated the provision of education for members then, two months later, called for a 'National Policy for Sanitarians'. Objectives and rules were formulated, but ultimately the society would remain under the control of the BUA. By September, the direction had shifted as the London Area had decided to form a 'New and Progressive Society' and in April 1927 the British Institute of Embalmers (BIE) was launched with W Oliver Nodes as its first president.

Education was the sole focus of the Institute's work.[41] Registration was also explored.[42] Although the BES and BIE ran in parallel by promoting embalmers until the 1950s (and the BES awarding a qualification until the decade prior), the Institute was undoubtedly the dominant organisation.

REGISTRATION AND THE NATIONAL ASSOCIATION OF FUNERAL DIRECTORS

As already stated, the desire to achieve registration was a clause in the founding objectives of the BUA. Indeed, the roots can be traced back further as, during the time of the BIU, an editorial in *TUJ* asked the question that if registration was achievable in the US, why not in Britain.[43] Henry Sherry raised the subject at BUA meetings in 1906 and 1907. However, there was dissent on the scope of registration: Thomas Kingston suggested complete burial reform instead of licensing, including the use of public mortuaries rather than through embalming, the latter being an idea endorsed by the Welsh embalming pioneer, Morgan R Morgan.[44] However, the degree of support for registration was questionable, with many believing that education first needed to be addressed.[45] Irrespective of this, the first draft of a Registration Bill was published in May 1922 but did not progress any further; it was followed two years later a more detailed document.[46] After a first reading by Thomas Groves, MP for Stratford on 5 March 1924, the Bill was withdrawn on 14 May, although the reason for this cannot be ascertained.[47]

Registration was discussed at the BUA conference in 1923 and, later the same year, a short questionnaire was sent to 1,500 MPs canvassing support.[48] In 1928–29, the BUA formed a registration committee and the matter resurfaced in earnest. *The BUA Monthly* stated, 'But in order to attain success, they must keep pegging away at the task of convincing their individual Members of Parliament at the need for Registration for the members of their calling.'[49] An editorial said that one objective of the Bill was to prevent undertakers touting; as Chapter 4 notes, this was a serious problem at the time among the estimated 10,000 undertakers in the country (4,000 being members of the BUA).[50]

However, trade union membership was an obstacle for the BUA and at the 1931 conference, the Manchester Centre put forward a resolution:

> 'That immediate steps be taken to remove the trade union status of the Association, the reason being that, in the opinion of the members, the progress of the Association is being retarded by the designation. In the alternative, in the event of such a step not being possible, that the Association cease to exist for a time and be re-organised under another name: for example, The Association of Funeral Directors.'[51]

Albert Cottridge believed that a registration claim would be built on education and knowledge:

> '... undertakers must show they perform a distinct social service; that this service is not a thing to be treated as a commodity bought and sold, but as something personal and to a large extent outside the range of ordinary commercial operations.' [52]

Furthermore, the BUA should '... be able to show that at least a substantial proportion of our members possess the necessary qualifications, and do, in fact, perform these special services.'[53] He also stated that, 'State registration is not a question of collecting many hundreds of pounds ... It is a question of deciding whether or not are they are convinced that the undertaker must be a sanitarian.' Cottridge further argued that registered funeral directors should inspect bodies and carry out tests for death.[54]

At a conference on registration held in February 1929, an advisory committee was formed where it was adopted that a 2s 6d capitation tax on all BUA members should fund the cause.[55] The

▼ In 1930, the BUA was honoured to have the famous Home Office pathologist Sir Bernard Spilsbury address their conference in Harrogate, at the Prince of Wales Hotel. He investigated numerous important cases such as Dr Crippen and the 'Brides in the Bath' murders. Sir Bernard is the tall, distinguished-looking figure holding his Homburg hat; he is standing with BUA president Harold [later Sir Harold] Kenyon on the left. The two standing together is not without irony: after his sudden death in December 1947, Sir Bernard was embalmed by Kenyon's (*BUA Monthly*)

▲ Lord Horder (1871–1955).
President of the Cremation
Society who brought the Funeral
Directors (Registration) Bill to
the House of Lords in June 1938
on behalf of the National Council
for the Disposition of the Dead

funeral industry was encouraged by the headway achieved by other occupations desirous of registration around this time: dentists and electricians, the Incorporated Secretaries' Association, and also architects.[56] However, the Harrogate undertaker, Ernest Swainson, believed that not all undertakers understood the implications of state registration, nor were all supportive.[57]

The most important attempt at registration was in the 1930s under the aegis of the president of the Cremation Society, Lord Horder. As the King's physician, he urged a linked between doctors and undertakers through the National Council for the Disposition of the Dead (NCDD).[59] The idea for an all-embracing council to discuss matters relating to disposal of the dead had been outlined at the BUA conference in 1932 by Dr H T Herring, the medical referee for the London Cremation Company.[60] This was endorsed by a fellow cremationist, Sir Peter Chalmers Mitchell, at a joint conference held in the same year, where the urgent necessity of setting up an advisory body of everyone directly or indirectly concerned in the disposal matters was emphasised.[61] The BUA, the National Association of Cemetery Superintendents (NACS) and the newly created Federation of British Cremation Authorities (FBCA) agreed to set up a central organisation. Lord Horder was hopeful that it would have a united voice:

> 'Closer liaison between doctor and undertaker is clearly needed if society is properly to be protected against crime and such co-operation will be rendered more easily and indeed obligatory if control is vested in this Council and it is recognised by the Government.'[62]

The secretary of the Cremation Society, P Herbert Jones, outlined the case for the NCDD at the second joint conference in 1933.[63] The following June, Lord Horder stated:

> 'It is the doctor who sees to it on behalf of those who remain that respect and reverence are paid to the departed friend, and finally it is he whose special knowledge is able to protect society by the sanitary disposal of the body, so that it in no way can ever be a danger to public health ... Close co-operation between those bodies is needed if the individual initiative, to which in the past this important subject has been left, is to be replaced by more organised effort.'[64]

At the State Registration Council meeting in 1934, its chairman, the Leeds undertaker, William Dodgson, reported:

> *'People were more and more realising that the members of the funeral trade should be highly skilled, whereas, at the moment, no qualifications were necessary and anyone could put up a sign and start in business as an undertaker. It was now being generally recognised that the disposal of the dead should be only in qualified hands. The National Council was willing to help this association to achieve its ends...'* [65]

▲ The lion rampant was adopted as the NAFD's logo

Comprising thirteen stake-holding organisations in the environment of the disposal of the dead, the objectives of the NCDD were:

- *The revision and codification of the laws governing the disposition of the dead*

- *The preservation of the land in the interest of the living*

- *Improvements of the status of those concerned with the disposition of the dead*

- *The safeguarding of public interest in all matters pertaining thereto.*[66]

At a conference held in 1934, P Herbert Jones (who would soon become secretary of the NCDD), said that overtures had already been made to the Home Office and Ministry of Health. He added:

> *'As long as he [the undertaker] has no status the public will not cease to regard the undertaker as a fit subject for music-hall jokes, side by side with mother-in-law'* [67]

P Herbert Jones believed that the keynote of any registration Bill should be education. Much discussion took place during 1935,[68] the year in which the BUA changed its name to the National Association of Funeral Directors. This took force on 1 November.[69] The NAFD objectives were summarised as 'Protection, Education, Qualification and Registration.'[70] It was the first major initiative to replace the description 'undertaker' with 'funeral director'.

By September the following year, a draft Bill for the 'State Registration of Funeral Directors' had been submitted.[71] One firm of funeral directors responded by publishing their own view on registration: all should be licensed, and there should be a three-year apprenticeship, testimonials and a written examination paper. There was a need to show that a '... better type of man is coming into the business ...'[72] However, it was not long before problems emerged. Ernest Swainson believed that '... we shall never get State Registration,' while the BUA recruited a solicitor to advise undertakers on how registration would affect them.[73] P Herbert Jones then requested need for evidence as to why funeral directors should be registered.[74] Tensions were also discernable when, at the 1937 joint conference of burial and cremation authorities, it was apparent that the NCDD's energies were solely being used to ensure the registration Bill became law and that no other issues were being explored, such as revision of disposal legislation.[75]

Undertakers were encouraged to muster support from local MPs, sanitary inspectors and medical officers of health. However, as one questioner at a BUA meeting asked, 'Will this opportunity be taken or will the profession be allowed to sink to a controlled "trade"?'[76] Undeterred, the NCDD pressed ahead by forming a consultative committee.[77] A paper on the subject by Lord Horder was also read at the NAFD conference.[78] In November, members of the NAFD were reassured concerning the Bill when the president of the association stated:

> '*I am fully convinced that any funeral director who is conducting his business today in a decent manner, or anyone who enters our vocation under the qualifying terms after the Act is in force and acts in a proper manner, has nothing to fear from the Bill.*'[79]

Lord Horder's Funeral Directors (Registration) Bill was presented to the House of Lords on 2 June 1938.[80] He briefly outlined that funeral directors who could conform to examination requirements would be registered under a statutory board, the chairman of which should be appointed by the Minister of Health. The Labour peer Lord Stragbolgi responded by saying that registration would lead to a monopoly situation, '... a closed corporation... ' and increase the cost of funerals.[81] Lord Eltisley, a Conservative, questioned the effect on part-time funeral directors in rural areas, while the Earl of Munster said that suitable safeguards were already in place through the remit of the Public Health Act 1936. Despite the Bill being defeated, Lord Horder thought the debate was 'unusually good', and continued to advance that the solution to reform was the registration of funeral directors.[82] The NAFD responded by saying they would welcome a Governmental Inquiry.'[83]

However, as Chapter 11 indicates, the association and the government were occupied by more pressing matters as the clouds of war gathered over Europe.

1. Parsons B (2005) 'From Institute to Association: The Formative Years of The British Undertakers' Association' *Funeral Director Monthly* Vol 88 No 11 pp8–9 and Vol 88 No 12 pp8–9

2. 'The British Institute of Undertakers; Conference at the Memorial Hall' (1898) *TUJ* October pp154–154a

3. 'The British Institute of Undertakers; Conference at the Memorial Hall' (1898) *TUJ* October pp154–154a

4. 'The British Institute of Undertakers; Conference at the Memorial Hall' (1898) *TUJ* October pp154–154a

5. 'The British Institute of Undertakers' (1899) *TUJ* December p164

6. 'Notes' (1900) *TUJ* July p76

7. 'Editorial: Shall we Embalm?' (1900) *TUJ* August pp97–98. See also 'Professor Sullivan's First English School of Embalming' (1900) *TUJ* October pp132–133

8. 'British Embalmers' Society' (1900) *TUJ* November p160

9. 'The British Embalmers Society' (1900) *TUJ* November p160

10. 'The British Embalmers' Society' (1901) *TUJ* May pp105–106

11. 'The British Institute of Undertakers' (1901) *TUJ* June p133

12. 'Editorial: The BIU' (1902) *TUJ* March pp75–76

13. 'Editorial: The BIU' (1902) *TUJ* March pp75

14. 'Editorial: The BIU' (1902) *TUJ* March p75

15. 'Interview with HA Sherry' (1904) *TUJ* March pp61–62

16. 'The North of England Funeral Undertakers' Association' (1903) *TUJ* February p ix

17. 'Biographical sketches by the Editor: Mr James C Broome' (1904) *TUJ* January pp15–16

18. 'Association of Metropolitan Undertakers' (1904) *TUJ* April p82

19. 'Proposed Undertakers' Convention' (1904) *TUJ* June p130

20. 'Liverpool Undertakers' Association' (1904) *TUJ* September p187

21. Foggart J (1904) 'Embalming and Sanitation: Their Value to an Up-To-Date Undertaker' *TUJ* October pp205–206, and Broome JC (1904) 'The Modern Undertaker' *TUJ* November pp233–234

22. Sherry HA (1904) 'The Needs of a National Association of Undertakers' *TUJ* October p212 and vii

23. Porter C (1904) 'The Advantages of Association' *TUJ* December pp259–263

24. 'Notes' (1905) *TUJ* February p25

25. *Hansard* 5 March 1924 Col 651 'Registration Bill for Undertakers. To provide for the registration of funeral undertakers and for the purposes connected herewith.'

26. Dr Alexander Wynter Blyth (1844–1921) was an eminent public health physician and president of a number of related organisations such as the Incorporated Society of Medical Officers of Health, and vice president of the Society of Public Analysts. He was also a barrister and author of books on hygiene and public health. His BUA paper was published as Wynter Blyth A (1905) 'The Hygiene of the Death Chamber' *TUJ* July pp156–161, and also published as a pamphlet by *TUJ*

27. 'Birmingham Organising' (1905) *TUJ* July pp191–192 and 'British Undertakers' Association' (1905) *TUJ* August p197. For a brief history of the BUA/NAFD see 'Brief Outline of the History and Objects of the NAFD' (1963) *NAFD Yearbook and Directory* pp11–15

28. Porter C (1905) 'How to Form Funeral Furnishers' Associations' *TUJ* July pp171–173

29. For biographical information about Sherry see, 'The Late Mr Henry Sherry. Father of the British Undertakers' Association' (1934) *TUJ* January pp3–4, and 'The Passing of a Pioneer' (1934) *BUA Monthly* January pp127–128. For biographical information about Hurry see also 'Interview with Councillor JR Hurry. Social Reformer and Undertaker' (1909) *TUJ* January pp13–15. Hurry was born on 9 March 1866 at Thornham Grove, Stratford. He died in March 1949. See 'Death of Mr JR Hurry' (1949) *FSJ* April p215 and James-Crook J (1949) 'Fifty Years of Service' *TNFD* May pp494–495. His daughter, Florence Delphine Hurry (19 August 1893–25 January 1974) was an embalmer and also national secretary of the National Association of Funeral Directors from 1946–1958. The funeral business closed around 1995

30. 'Report of the Committee on Amalgamation of the British Undertakers' Association and the British Embalmers' Association' (1915) *TUJ* July pp182–183; 'Working out the Amalgamation' (1915) *TUJ* August pp209–212; 'The Proposed Amalgamation' (1915) *TUJ* October p273; 'Amalgamation of the BES and BUA' (1915) *TUJ* November pp299–303; 'The News Association' (1915) *TUJ* December pp337–338; 'British Undertakers' Association' (1916) *TUJ* April pp96–99 and May p126

31. See 'London Funeral Carriage Proprietors' Association' (1916) *TUJ* September p253, October pp275–276, November p302 and December p330. See also 'London Funeral Carriage Proprietors Association' (1916) *TUJ* October p275 and November p302 and p253

32. Nodes HK 'A Minimum Charge for Carriages' letter (1916) *TUJ* October p273

33. 'LFCPA and the BUA' (1917) *TUJ* April p93

34. 'Trade Meeting at Manchester' (1918) *TUJ* April p86. Registration was granted on 28 December 1917 (Number 1665T)

35. 'British Undertakers' Assn. Bolton and District Centre' (1917) *TUJ* May pp121–122

36. 'British Funeral Workers' Association' (1917) *TUJ* September p234, and October p261. 'The National Union of Funeral and Cemetery Workers. A Review of its Growth, Scope and Objects' (1950) *FSJ* June pp325–326 and 'NUFCW Celebrates Golden Birthday' (1967) *FSJ* September pp477–478. See also 'NUFSO joins FTAT' (1978) *FSJ* July p276, and 'A Change of Name' (1969) *FSJ* July pp367–368

37. 'Interview with Mr Thomas Kingston' (1918) *TUJ* January pp19–20; 'Interview with TW Kingston of Stratford' (1907) *TUJ* December pp295–296 and (1908) *TUJTUJ TUJ*January pp17–18. See also 'Co-operation in the Funeral World' (1920) *TUJ* February p41 and 'Interview with Mr TW Kingston; Organiser of the Funeral Workers' Association' (1925) January pp17-18

38. 'The British Undertakers' Assn. London Centre' (1918) *TUJ* April p93

39. 'Editorial Notes: The Landaulette Question' (1924) *BUA Monthly* December p164 and 'Agreement Reached on Amalgamation' (1924) *BUA Monthly* December pp173–175

40. 'Embalming in England' (1930) *TUJ* August pp258

41. Parsons B (2007) 'Preservation and Professionalisation: Tracing the Development of the British Institute of Embalmers' *The Embalmer* Vol 50 No 3 pp16–20

42. 'Embalmers and Registration. Case for the Draft Bill Explained' (1937) *TUFDJ* June pp185–187 and 192–198

43. 'Editorial: Registration' (1898) *TUJ* May p76

44. Sherry HA (1906) 'The Need for a Licence' *TUJ* September pp232–233. See also 'Editorial Licence versus "Licence"' (1906) *TUJ* September pp230–231, and Sherry HS (1907) 'The Licensing of Undertakers' *TUJ* February pp48–50.

'Editorial: 'Registration' (1916) *TUJ* September p249–250. Morgan MR (1916) 'Registration and the Need of the Moment' *TUJ* December p326, and 'The Question of Licence' (1907) *TUJ* November pp269–271

45. Mountfield A (1922) 'Registration: First Impressions' *TUJ* June p181

46. 'Registration' (1922) *BUA Monthly* May pp250–251. See also Cole P (1921) 'Incorporation and Registration' *BUA Monthly* June pp338–340; 'A Proposed Bill' (1923) *BUA Monthly* December p525; and 'State Registration: First Draft of the Bill to Provide for the Registration of Persons Undertaking the Disposal of the Dead' (1924) *BUA Monthly* June pp687–690

47. *Hansard* 5 March 1924 Col 651 'Registration Bill for Undertakers. To provide for the registration of funeral undertakers and for the purposes connected herewith.' *Hansard* 14 May 1924 Col 2200. See also, 'Thoughts on the Registration Bill' (1925) *TUJ* May pp169–170; and Nodes HK (1925) 'State Registration' *BUA Monthly* August pp 36–39. See also, 'The Woman's Page' (1925) *BUA Monthly* May p300

48. 'The Case of Registration' (1923) *TUJ* August pp291–292. See also Nodes HK (1923) 'State Registration' *BUA Monthly* July pp388–389, and Nodes HK (1923) 'State Registration' BUA Monthly December pp521–526 and pp525–526. 'The MP's Questionnaire' (1923) *BUA Monthly* December pp531–532. See also Kingston TW 'The State Registration of Undertakers: Some Practical Difficulties' (1923) *TUJ* October p351 and 'The Why and How of Registration' (1923) *TUJ* November p389

49. 'Registration' (1928) *BUA Monthly* August p49

50. Editorial Notes: 'State Registration: What Ththis Means' (1928) *BUA Monthly* October pp99–100

51. 'Trade Union Status' (1931) *BUA Monthly* June p267

52. Cottridge AJE (1928) 'State Registration' *BUA Monthly* December p129

53. Cottridge AJE (1928) 'State Registration' *BUA Monthly* December p130. See also Editorial Notes: 'Education and State Registration' (1928) *BUA Monthly* December p125

54. Cottridge AJE (1929) 'Preparing the Way' *BUA Monthly* March pp196–198. Albert Cottridge repeated this in a later paper: see Cottridge AJE (1936) 'State Registration for Funeral Directors' *TUFDJ* September pp301–302. See also 'The Demon Undertakers. A Hound with a Hearse' (1929) *BUA Monthly* January pp152–153

55. 'State Registration Conference. Joint Meeting in London' (1930) *BUA Monthly* March pp 200–211. See also 'State Registration' (1929) *BUA Monthly* March p229

56. Downer A (1930) 'Some Registration Considerations. Is the Registration of the Secretary a Feasible Proposition?' *BUA Monthly* February pp170–171. See also 'State Registration Conference' (1929) *BUA Monthly* May pp259–263, and McArthur Butler A (1931) 'The Registration of Architects in Great Britain' *BUA Monthly* September p67–69

57. Swainson WE (1931) 'A Review of Association Activities with Special References to Education' *BUA Monthly* December p122–124. See also 'What Registration – What BUA Members should do' (1932) *BUA Monthly* May pp254–255; 'State Registration: What are the Prospects?' (1932) *TUJ* March pp95–96. See also 'Letters to the Editor: The National Council for the Disposition of the Dead' (1934) *TUJ* October p343; 'Letters to the Editor: State Registration: Some Queries' and 'Lord Horder and the Undertakers' (1935) *TUJ* August pp264–265; 'Letters to the Editor: The Registration Bill: A Plea for Caution' (1935) *TUJ* October p332; 'Letters to the Editor: Lord Horder and the Funeral Directors' (1936) *TUFDJ* January p52

58. Spilsbury B (1930) 'The Sanitary Care of the Dead' *TUJ* July pp237–238. See also 'Sir Bernard Spilsbury Addresses the Conference' (1930) *BUA Monthly* August pp28–33

59. 'Important Medical Support of the Undertaker. Lord Horder's Statement' (1933) *BUA Monthly* August p41. For a full account of the NCDD see Jupp PC (2006) *From Dust to Ashes. Cremation and the British Way of Death* Basingstoke: Palgrave Macmillan pp114–119, and Jupp P (2008) 'The Council for the Disposition of the Dead 1931–1939' *FSJ* July pp103–106

60. Herring HT (1932) 'Are we to let the dead bury the dead?' *BUA Monthly* August pp45–48; 'The Cemetery and Cremation Authorities confer with the BUA (1932) *TUJ* March p85. See also 'National Council for the Disposition of the Dead (1933) *TUJ* March pp91–92 and 'Editorial: A National Council' (1933) *TUJ* March pp89–90

61. Mitchell PC (1932) 'The Disposal of the Dead' *TUJ* August pp243–245

62. Horder, Lord (1934) 'The Disposition of the Dead. The Need for Co-operation' *BUA Monthly* June pp238–239

63. Jones HP (1933) 'The Disposition of the Dead – The Case for the National Council' *TUJ* November pp367–369

64. Horder L (1934) 'The Disposition of the Dead. The Need for Co-operation' *BUA Monthly* June pp238–239

65. 'State Registration Council' (1934) *BUA Monthly* June p241

66. 'The National Council for the Disposition of the Dead' (1934) *BUA Monthly* July p3. See also Horder, Lord (1936) 'The Public Health Aspect of the Disposition of the Dead' *TUJ* November pp357–358

67. Jones PH (1934) 'The Way to Sate Registration' *BUA Monthly* September pp74–80 and Jones PH (1934) 'State Registration for Undertakers' *TUJ* September pp293–296, and 'The National Council for the Disposition of the Dead' (1934) *TUJ* November p399

68. 'The National Council for the Disposition of the Dead' (1935) *BUA Monthly* February p162; 'National Council for the Disposition of the Dead' (1935) *TUJ* February pp45–48; 'The National Council for the Disposition of the Dead' (1935) *BUA Monthly* June p246; 'The National Council for the Disposition of the Dead' (1935) *TUJ* May p159 and 'The National Council for the Disposition of the Dead' (1935) *TUJ* December pp387–388

69. 'Editorial: The Reorganization of the BUA' (1935) *TUJ* February pp55–56; 'Editorial: What's in a Name?' (1935) *TUJ* October pp335–336 and Perritt JD (1935) 'The National Association of Funeral Directors. The National President on the New Constitution' *TUJ* November p371

70. 'National Association of Funeral Directors' (1935) *BUA Monthly* August p30 and 'The National Association of Funeral Directors' (1935) *TUFDJ* December p389

71. 'The Disposition of the Dead' (1935) *TNFD* December pp125–126, and Jones PH (1935) 'The New Constitution of the National Council for the Disposition of the Dead' *TUJ* November pp375–379. See also 'The Registration Bill' (1935) *TUJ* July pp233–235. For copy of the draft Bill see 'The Draft Bill to provide for the registration of funeral directors and for purposes connected therewith' (1935) *TNFD* December 1935 pp128–132 and *TUJ* December pp393–396. See also 'Editorial: The Bill for the State Registration of Funeral Directors' (1936) *TUFDJ* January pp21–22 and 'Editorial Notes' (1936) *TNFD* September p74

72. 'Registration and Funeral Directors' (1936) *TUFDJ* October p330

73. Swainson WE (1936) 'Propaganda' *TNFD* August pp41–45; and Littlewood SCT (1936) 'State Registration' (1936) *TNFD* November pp109–110

74. Jones PH (1936) 'The Need of Evidence' *TNFD* December p124 and 132

75. Jones PH (1937) 'The Council for the Disposition of the Dead. Retrospect and Prospect' *TNFD* January pp153–154

76. 'Impressions of an Observer' (1937) *TNFD* January p164. See also Jones PH (1937) 'General Principles of Registration' *TNFD* March pp213–214 and Jones PH (1937) 'The Effects of Registration on Other Professions' *TNFD* May p270 and 272

77. 'State Registration. Parliamentary action to be taken without delay' (1937) *TNFD* May p261

78. Horder, Lord (1937) 'Planning for the Disposal of the Dead' *TNFD* August pp59–60, and *TUJFDJ* (1937) August pp271–272. The paper was read in Lord Horder's absence

79. 'The Registration Bill' (1937) *TNFD* November p122

80. 'Funeral Directors (Registration) Bill' (1938) House of Lords Debates 2 June Cols 853–866, and 'Registration of Undertakers' (1938) *TUFDJ* June pp213–214. See also 'Registration Bill Prospects' (1938) *TUFDJ* May p179, and Jupp PC (2006) *From Dust to Ashes. Cremation and the British Way of Death* Basingstoke: Palgrave Macmillan pp114–119

81. The Bill was objected to by Lord Strabolgi. Curiously he promoted a bill to register hairdressers in 1937. See 'Notes' (1937) *TUFDJ* July p206

82. 'Lord Horder's Report' (1938) *TNFD* August p54. See also 'Editorial: The Registration Bill' (1939) *TUFDJ* June pp211–212. See also Porter C (1938) 'Funeral Directors (Registration) Bill' *TUFDJ* September p317, and Horder, Lord (1941) 'A Neglected Public Duty. The Problem of Burial *TNFD* April p326

83. 'Funeral Directors (Registration) Bill' (1939) *TNFD* May p432

CHAPTER 3
THE EMBALMER AT WORK: SALISBURY 1906

On the evening of Saturday 30 June 1906, the London and South Western Railway's (LSWR) American Line boat train left Plymouth bound for Waterloo.[1] It was a weekly trip scheduled to arrive in London four hours and twenty minutes later. The train carried 43 passengers who had arrived in England on board the SS *New York* liner. However, the train failed to reach its destination as, just before two o'clock the following morning, it derailed on a sharp curve at the east end of Salisbury station. Although there was a restriction of 15mph, the train was travelling nearly four times this speed. The locomotive left the rails and collided with a train of empty milk vans. The damage caused by the crash was as follows:

> 'No. 421 [the engine] came to rest on its right-hand side with the tender jack-knifed into the cab, but with surprisingly light damage. The leading three vehicles were effectively completely destroyed, with the fourth vehicle sustaining heavy damage to one end and with a complete side torn away. The rear kitchen brake sustained minor damage and came to rest with its rearmost wheels still on the track. Of the 21 vans on the milk empties train, five were destroyed, with five others sustaining varying degrees of damage.'[2]

The crash caused the death of 24 passengers, in addition to the driver and fireman of the boat train and the guard and fireman of the goods train. Many others sustained terrible injuries. An emergency operation was immediately enacted as medical staff, police, railway employees, postal workers from the nearby sorting office and local residents woken by the noise of the impact made their way to the station. What they faced was described in *The Salisbury and Winchester Journal*:

> 'The force of the impact when the smash occurred was terrible and it was heard all over the city. A scene of devastation met the eyes of those who rushed to the station. In the prevailing gloom could be seen engines and carriages lying in a shapeless heap, whilst strewn about the platform and permanent way were axle

▼ Part of the luggage van overhanging Fisherton Street Bridge in Salisbury

▲ The scene at Salisbury on Sunday morning before the wreckage had been removed

BOAT TRAIN DISASTER AT SALISBURY.

boxes, milk cans, carriage panels, and cushions torn to shreds ... Amidst the wreckage were unfortunate passengers, terribly lacerated, with torn and broken limbs, writhing in agony, and by their side lay the motionless bodies of those with whom the hand of Fate had perhaps been more merciful in releasing them speedily from their suffering.'[3]

In respect of the recovery of the dead and injured, *The Salisbury and Winchester Journal* noted:

'The work of the rescue was rendered the more difficult owing to the darkness, but helpers streamed into the station, and by the aid of lamps and lanterns did their best to relieve the injury of the suffering. The waiting-rooms at the station were hastily prepared for use as surgeries and mortuaries, and in a very short time thirteen bodies lay side by side in the ladies' waiting room ... The work of extricating the dead, however, was not completed until Sunday afternoon. A number of bodies could not be extricated owing to the weight of splintered wood, twisted iron, beneath which they were buried until the arrival of the breakdown gang from Nine Elms and Eastleigh. It was nearly noon when seven bodies were taken from the wreckage of the second carriage, and some two hours later those of the driver and fireman of the express were recovered from beneath their engine.'[4]

As soon as the bodies had been removed, re-railing and clearance of the tracks commenced. A Board of Trade inquiry was immediately ordered into what was the LSWR's worst accident to date.[5] Major J W Pringle, Inspecting Officer of Railways, started his investigation on the day of the accident. However, before bodies could be transported, an inquest had to be opened. Conducted by the coroner for the city of Salisbury, Mr S Buchannan Smith, this took place on the afternoon of Monday 2 July. As soon as the jury was sworn in, they were taken to the hospital and railway station to view the bodies, as was the practice at the time. When they returned to the court, statements were heard from those who had confirmed identification of the bodies, before the inquest was adjourned until Monday 16 July.

THE EMBALMERS ARE SUMMONED

The majority of those killed were Americans and it would appear that, as soon as news of the accident reached overseas, instructions were received to repatriate the bodies for burial. It was therefore essential that embalming was carried out. What happened next is described in *TUJ*:

▲ Members of the embalming team, L to R: James Goulborn, John Dacombe, Herbert Hollick Kenyon, Harold Vaughan Kenyon, Harry Woolcott

> 'Mr Dacombe, head of the firm of John Beeston & Co [funeral directors] of Southampton, who had received instructions to make the necessary arrangements with regard to ten of the victims, at once realized that it would be quite beyond the power of any one man to undertake the embalmment of such a number in the necessarily limited time, and therefore telegraphed at once to Mr Hanshue, the well-known manager of the Buckhout and Breed Chemical Company, for the services of two qualified embalmers. Messrs. Kenyon, of Edgware Road had already received instructions in respect of six cases; it was only therefore necessary for Mr Hanshue to reply that Messrs HH and HV Kenyon, their Mr Woolcott, and Mr James Goulborn were on their way to Salisbury. Mr Edwin, of Guildford – who recently took a course of instruction in embalming – was also present, and rendered valuable assistance.'[6]

It was somewhat fortuitous that about six years prior to this accident, English undertakers had been given the opportunity to receive embalming training from Professors Renouard and Sullivan. Harold Kenyon, son of the founder of J H Kenyon, was among the 30 undertakers who completed Renouard's class, while Herbert Hollick Kenyon (who had married into the Kenyon family and adopted the name) attended the course run by Professor Sullivan. Having joined the firm in May 1893 aged 14 years, Harry Woolcott was manager of J H Kenyon's branch at

12 Church Street, Kensington, but was not an embalmer.[7] James Goulborn had a funeral-directing business at 43 Greek Street in Soho and by 1904 was teaching at embalming training sessions held under the auspices of the London branch of the British Embalmers' Society.[8] Mr Walter James Dacombe's business was based at 123 St Mary Street in Southampton. Being near the port, he was probably no stranger to repatriations.[9] No details about Mr Edwin can be traced.

When interviewed about the events for an article published in *TUJ* in October, Herbert Hollick Kenyon recalled:

> 'The accident, as you know, took place at 2am on Sunday, July 1, but we were unable to begin work until Monday noon, thirty-six hours later, owing to the delay in obtaining the permission of a coroner and chief constable. Under these circumstances, in connection with the exceeding hot weather prevailing at the time, pronounced signs of decomposition were in evidence when the bodies were turned over to us. The railroad officials were naturally anxious for us to finish the work at the earliest possible moment, and observing their wishes in that direction we worked incessantly from Monday noon until Tuesday evening, when the work was practically completed, and fifteen of the bodies were placed in the leaden inner caskets ready for removal.'[10]

Again, *TUJ* gives an insight as to the extent of the injuries and the embalming treatment:

> 'In many cases the injuries were very severe, among the most seriously mutilated being a lady who was literally torn in halves and a gentleman whose lower limbs were mangled beyond recognition. Needless to say these cases received specially careful attention, fully warranted by the results, for in neither would it have been possible for anyone subsequently viewing the remains to have imagined the terrible disfigurement to which they had been subjected. Every case called for careful thought, the arterial system being frequently so deranged that a complete circulation of the fluid by ordinary means was an impossibility; in two case, for instance, the

◀ A photo taken from the roof of the Great Western Railway terminus building at Salisbury showing the locomotive cranes removing wreckage

▲ The scene early the next morning before the cranes had removed any of the wreckage

▼ A close-up view of the damage to the express train, engine no. 421

heads were crushed out of all human semblance, apart from other severe injuries. From a technical point of view a very interesting case was that of a gentleman who externally showed but little sign of injury, but the injection by means of the brachial artery of 128 ounces of fluid clearly showed that the circulation was disorganized, as no evidence of its presence in the tissues could be detected. There is no doubt that this was due to injury to one or more important arteries in the thorax, many ribs being broken.'[11]

Herbert Hollick Kenyon also added:

'We embalmed as thoroughly as such unfavourable conditions would permit. Our main reliance, however, was the use of the hypodermic needle, with which we went over the entire surfaces, making injections therewith close enough together to ensure thorough embalming. In the cases where the head was considerably damaged, we used hardening compound in generous quantities and injected fluid again by the hypodermic process.'[12]

Although the information above states that fifteen cases were embalmed, examination of Kenyon's registers shows that it was actually 17.

From the accounts of the accident, it can be surmised that the embalming took place in one of the waiting rooms on Salisbury station, and also at the hospital. As one waiting room served as a makeshift mortuary, it would be reasonable to assume that this became a makeshift embalming theatre. It is not known what fluids were used, although during 1906 *TUJ* carried advertisements for 'OK & Special' embalming fluids supplied by the Buckhout and Breed Chemical Co of London; fluids could also be obtained from the wholesale supplier, Dottridge Bros.

Although the embalmers commenced their work immediately after the coroner and chief constable had given their permission, nowhere in accounts of the incident or inquest does it mention that the bodies were subject to a post mortem examination. If this had been carried out, medical evidence would have been presented by a pathologist at the inquest. In view of the circumstances of the accident and the extent of the injuries, it is likely that the coroner considered such an examination to be unnecessary.

Worthy of mention is the statement above concerning, '... the exceeding hot weather prevailing at the time.'[13] Research indicates that the temperature in Salisbury on the day of the accident was around mid 60°F. The weather report in *The Times* states that, 'Over the general part of England the weather yesterday [1 July] was fair and dry, but distinctly cool.'[14] On Tuesday – the second day of embalming – the maximum air temperature in England was 71°F, with maximum sunshine of 6.6 hours being recorded in Westminster.[15] It is surprising that with such modest temperatures the weather was described as 'hot'.

After the embalming had been concluded, the bodies were transferred from Salisbury to await repatriation. Some, however, were destined for burial in England, such as the four railway employees and one of the passengers, the vicar of St Thomas' Church, Toronto. After a service in Salisbury Cathedral, he was buried in Devizes Road Cemetery.

THE REPATRIATIONS

As noted above, Kenyon's had been engaged to repatriate a number of the bodies: their funeral registers detail the type of coffin, logistical arrangements, other services provided and costs. The coffins would have been prepared at either the Edgware Road or Shelton Street (Paddington) branches, and then taken by horse-drawn hearse to Waterloo station where they would have been conveyed by train to Salisbury. Kenyon's staff would have placed the deceased in the coffins before making the return journey to the firm's private mortuary chapel in Paddington. Under the Kenyon diary entry for Monday, 2 July 1906, there are three entries describing the services provided:

- *To embalming the remains*
- *Pine shell, lined 2s*
- *SS & P [side sheets and pillow]*
- *Lead coffin*
- *Conveyance and men to Waterloo and Salisbury*
- *Hearse and pair, Waterloo to Mortuary Chapel*
- *Driver, 4 men, pall*
- *Mahogany case, moulded lid and plinth French polished 4 pairs silver-plated fittings*
- *Silver-plated [name] plate*
- *Conveyance and men*
- *Packing case.*

▲ The coffins containing Driver Robins and Fireman Gauld being placed on a special train at Salisbury

The size of the coffin, details for the name plate (full names and date of death; some also had the date of birth) and names and address of the person receiving the funeral account are also given in the margin of the diary. In addition, there is a further entry for the 'Temporary Embalmment of Remains' and provision of a 'garment' for one of the casualties. This 30-year-old male was a butler to one of the families on the train. He was from Newcastle and his body was taken there for burial.[16] A charge of £5 5s was made for the embalming and 3s 6d for the 'garment' (a shroud). In respect of the former expense, this was commensurate with the embalming charge usually made by the firm. A Kenyon advertisement in *The Kensington News and West London Times* in October 1899 states the cost of embalming ' ... varies between £5 5s and £15 5s'.

Supplying the coffins and releasing staff to embalm had to be managed alongside Kenyon's domestic work. Fortunately, the records indicate that it was not a busy time. On Monday 2 July, instructions for two funerals were received but there was only one funeral scheduled: a burial at Brompton Cemetery requiring a brougham and pair (of horses), two drivers and six men. In addition, a hearse with a pair of horses, driver and six men were sent to Victoria station to meet the 8.50 p.m. train containing a coffin sent from Brighton-based funeral directors, Attree and Kent, which had to be transferred across central London to Euston station for an onward journey.

On Tuesday 3 July, Kenyon's transferred a further three pine shells, lead coffins and mahogany cases to Salisbury. One remained at Salisbury. However, along with the three brought down on Monday, *The Times* reveals that five coffins were placed on the 6.28 train to Waterloo on Tuesday evening.

The register also lists all the names of those who were embalmed but were not repatriated by Kenyon's. The firm provided a 'pine shell lined with side sheets, ruffles and pillow' that was then placed in a lead coffin. Including the transport to Salisbury, Kenyon's issued an invoice for £10 4s 6d to Beeston & Co, who would have been responsible for the repatriation of one person. The coffins were taken to Southampton on the 6.55 p.m. train and remained in the docks until Saturday 14 July before they were transported to New York.

On Wednesday 4 July, the entry for the last embalment appears in the register. This is charged at a higher rate of £20. A note also heads the page: 'Mr R Gray of *Morning Post* wants us to inform him of time and place of memorial service for the Americans.' This was arranged by the Dean and Chapter of St Paul's Cathedral for 10 July.[17]

The first repatriation took place on Friday 6 July. Kenyon's foreman and a plumber travelled by the 2.55 p.m. train from Waterloo to Salisbury to seal the coffin. They then took the 6.55 train to Southampton where the coffin was taken on board the North German Lloyd liner *Kaiser Wilhelm II* destined for New York.[18]

The next repatriation occurred on Tuesday 10 July when one coffin was taken from Kenyon's premises to Euston station for the 12.25 morning train to Liverpool. The coffin was then loaded on the SS *Teutonic* bound for New York. Finally, the remaining four coffins were taken to Euston station for the journey to Liverpool on the evening of Monday 16 July. The Cunard steam ship *Carmania* departed the following day, along with Herbert Hollick Kenyon, who had the unusual task of accompanying the coffins to New York, docking on 24 July.

Commenting on the journey, Kenyon noted:

> 'On the trip across the ocean … the bodies were accommodated in the steerage, with one exception, in which case a special space was arranged for. The sailors … were not in the least nervous, and stories relating to the "superstitions" in regards to dead bodies aboard ship appear to be pure fiction.'[19]

In response to a question about the condition of the bodies, he replied:

> 'We had every reason to be satisfied with them. The coffins were not opened, the relatives not desiring they should be, considering the mutilation some of them had undergone. But there was no trace of decomposition and the tissue had become firm and

*hard, and I am satisfied that the embalming was all
that could be desired.'[20]*

Just before the bodies were transferred to Southampton, they had
in fact been viewed. At the time it was noted that:

*'One could not but be struck ... with the life-like
appearance [of] the face of each presented, so unlike
what we usually see in the dead, and this though
ten days had elapsed since the terrible accident at
Salisbury. What is more, notwithstanding this delay
in burial, there was not perceptible the slightest trace
of odour.'[21]*

Kenyon's final account is detailed in the registers:

*10 p.m. train Euston to Liverpool for shipment on
"Carmania" to New York.*
- *Bus and pair*
- *Conveyance*
- *drivers 6 men*
- *Rail charges, fares, etc*
- *Shipping charges*
- *HHK to New York and half passage money*
- *Consular fees*
- *Floral decorations.*

It would appear that the cost of Herbert Hollick Kenyon
accompanying the bodies was divided between two of the four
invoices. This accounts for two being charged at £101 16s 6d, and
the others at £135 8s 6d.

While abroad, Herbert Hollick Kenyon took the opportunity
of visiting funeral furnishers' premises in New York and also in
Canada. In an article in *TUJ*, it was reported that he had been
very impressed with the facilities. He said:

*'When you entered an American undertaker's
establishment you were not confronted by a workshop
or by a show of funeral appurtenances, but simply with
a business office, clean, neat, and nicely furnished,
with showrooms attached, and a mortuary chapel
in which a service might be held, or embalming
operations carried on.'[22]*

He noted that workshops were not needed as finished caskets were purchased from manufacturers. Kenyon also recommended that English undertakers visit the States and Canada, ' ... and thus see American methods. It would do good.'[23]

MAJOR PRINGLE'S CONCLUSION

After sitting with the coroner as an assessor (to provide technical information) and conducting his own enquiry, Major Pringle's official report into the accident was published on 31 July 1906. The excessive speed of the train was the official cause. Major Pringle stated:

> ' ... it is inexplicable how a driver with the experience of Robins could have been apparently so reckless as to travel at such a high speed on this section of the line.'[24]

But as the railway historian Norman Pattenden notes, it is clear that, 'Major Pringle had some element of doubt in his mind; after all, everything had been normal other than for Driver Robins not applying the brake ... The "Why?" was to remain in Major Pringle's report.'[25]

Immediately after the accident, rumours about its cause were rife: competition between railway companies to complete the journey the quickest; the driver trying to impress his colleagues; a bribe by a passenger to reach Waterloo before time; and a misjudgment about the centre of gravity of the engine were all voiced.

After researching Driver Robins's work and rest over the days immediately prior to the accident, Pattenden suggests that he may have been experiencing a 'micro-sleep' as a result of his changing shift patterns. A short period of sleep on the footplate as the train approached Salisbury may well have prevented him from sounding the whistle and failing to apply the brakes. This is a well-argued explanation, but is still conjectural. Indeed, it is likely that the reason will never be established. However, as Pattenden concludes: 'At the very least, Driver Robins deserves to be given the benefit of the doubt.'[26]

1. For further accounts see: Faulkner JN and Williams RA (1988) *The LSWR in the 20th Century* Newton Abbot: David & Charles pp178–181; Rolt LTC (1982) *Red for Danger* Fourth Edition Newton Abbot: David & Charles pp166–168; Bradley DL (1986) *LSWR Locomotives: The Drummond Classes* Didcot: Wild Swan Publications pp 60–61; Moody JB and Fleming G (2006) *The Great Salisbury Train Disaster Centenary 1906–2006: Voices from the Boat Train* Salisbury: Timezone Publishing; Pigott N (2006) 'Salisbury: still a mystery 100 years on' (2006) *The Railway Magazine* July pp28–31 and 'Salisbury: Did the driver go into a micro-sleep?' (2006) *The Railway Magazine* October p26

2. Pattenden N (2001) *Salisbury 1906: An Answer to the Enigma* Swindon: The South Western Circle p5

3. *The Salisbury and Winchester Journal, and General Advertiser* 7 July 1906 p6

4. Ibid

5. Faulkner JN and Williams RA (1988) *The LSWR in the 20th Century* Newton Abbot: David & Charles p179

6. 'The Salisbury Railway Disaster' (1906) *TUJ* July p175

7. 'Seventy Years of Funerals!' *FSJ* July 1961 pp317–372 (Information received from the late Des Henley.)

8. 'Interview with Mr James Goulborn of Greek Street, Soho, London' (1904) *TUJ* November pp237–238

9. A former mayor of Southampton, Walter J Dacombe, died on 15 September 1921

10. 'Interview with Mr HH Kenyon' *TUJ* (1906) October pp267–268

11. 'The Salisbury Railway Disaster' *TUJ* (1906) 15 July p175

12. 'Interview with Mr HH Kenyon' *TUJ* (1906) October pp267–268

13. Ibid

14. *The Times* 2 July 1906

15. *The Times* 4 July 1906

16. *The Times* 4 July 1906

17. *The Times* 11 July 1906

18. *The Times* 5 July 1906

19. 'Interview with Mr HH Kenyon' *TUJ* (1906) October pp267–268

20. Ibid

21. 'The Salisbury Railway Disaster' (1906) *TUJ* July p175

22. 'Interview with Mr HH Kenyon' (1906) *TUJ* October pp267–268

23. Ibid

24. *Accident Report* Board of Trade 1906

25. Pattenden N (2001) *Salisbury 1906: An Answer to the Enigma* Swindon: The South Western Circle p69

26. Ibid p72.

CHAPTER 4
FUNERAL SERVICE DURING WWI

Although commencing in the Summer of 1914, the Great War made little impact on the funeral industry until 1916, when the Military Service Act came into force on 27 January. Legislation conscripted every British male who, on 15 August 1915, was aged between 19 and 41 and was unmarried or a widower without dependent children.

The requirement for all able-bodied men had an almost immediate impact on both undertakers and cemeteries. In March 1916, the BUA secretary and an Islington-based undertaker, William Spicer, wrote to the Board of Trade about the shortage of labour, only to be met with curt rejection: 'We are unable to recommend the exemption of undertakers or their employees.'[1] This was followed by the equally terse: 'I am to add that no useful purpose would be served by a deputation.'[2] Later in the year, the newly formed National Association of Cemetery Superintendents (NACS) would also seek an exemption, only to be met with the same response.[3]

Although few records can be identified, the shortage of staff made a considerable impact upon firms. An employee list was maintained in J H Kenyon's the funeral registers and at least ten members of staff were listed as being on 'active service' in 1914.[4]

The owners of businesses applied to tribunals for exemptions to retain their key staff. These were set up in most main towns to hear applications from employers and parents; the *TUJ* published summaries of the cases, as these examples indicate from April, May and July 1916:

> 'A coffin dealer applied before the Edinburgh Tribunal for exemption. On his behalf an agent stated that eleven years ago he started a small joiners business. Since then the business had developed considerably, and appellant was now a large dealer in coffins, and supplied the majority of the undertakers in the city. Owing to his expert knowledge, he had been able to make "a special corner in cheap coffins." (Laughter.) No one else in the city appeared to have developed his line. The Sheriff remarked that this was one of these special cases that the regulations submit to the consideration of the Tribunals. It was a very exceptional employment,

*and certainly supplied a need of the community.
It seemed to him that the man should be exempted.
Conditional exemption was granted.*[5]

*'An undertaker appealed at the Poplar Local Military
Tribunal for exemption for his son, and stated that he
was unable to replace him "as nine out of every ten
men won't touch a corpse." He added that he had only
one other man and his son; but he had two sons in
the army. The Court understood the difficult position
in which the undertaking trade was placed owing to
the scarcity of men; and the work of burying the dead
must be attended to. – He was granted six months'
conditional exemption.*[6]

*'The Deputy Town Clerk of Greenwich applied for
the total exemption of six employees of the Borough
Council Cemeteries at Shooter's Hill and Charlton,
they being indispensable as gravediggers. He said that
it had been impossible to get excavators, even though
they had applied at the Labour Bureau. In fact, they
had had to transfer men from other departments
of the Council for cemetery work. The number of
interments had recently increased. Three of the men
were exempted on condition that they continued the
occupation of gravediggers, and the other three were
given three months' exemption.*[7]

*'An undertaker of Stratford put forward a claim on
behalf of his son, who looked after one of his shops,
which must be closed if his son went. He said there
was a willingness on the part of all "to do their bit"
but people would insist on dying, and must be buried.
Three months' exemption was granted.*[8]

*'At the Southwark Tribunal, an undertaker, of London
Road, applied for further time for his son, single, aged
21. He is a coffin maker, and had already received three
months' exemption. If the son was called up he would
have to refuse to conduct any more funerals. He had
been forced to close one shop in Brook Street, when his
last man went. For twelve months he had worked the
business with his son. When there were two funerals*

he conducted one, and his son acted as substitute at the other. The Mayor: How do you go on for bearers? We hire them from other undertakers. He added that on one occasion his firm was so pressed that four days elapsed before they could make a coffin for a corpse. – Two months' extension was granted.'[9]

Later in the year, one tribunal suggested greater co-operation between undertakers:

'The undertakers in Coventry appeared before the Tribunal asking for exemption for their coffin makers, seven in number, two being single and five married. Six firms were concerned in the claim, and they were represented by Mr William Maddocks. In the course of the hearing it transpired that one firm had conducted 513 funerals during the last six months and another firm 332 funerals in the same period, and that fifty per cent of the funerals were their own orders.

Mr Maddocks explained that these firms supplied hearses and coaches for smaller undertakers in the city. In normal times a stock of fifty to a hundred coffins was kept, but at the present time there was no reserve, and one manufacturer had to help another in an emergency. It was impossible to buy coffins wholesale, or to get fresh coffin makers.

Mr Farren: Some of us will have to be content to be buried without a coffin.

Mr Maddocks: As the law stands today, a man must be buried in a coffin.

The Chairman suggested that it would be possible for the men in the trade to get together and arrange a scheme so that the coffin makers should work together and provide all the undertakers.

Mr Farren: We suggested such a procedure in the milk trade and I see no reason why this particular trade cannot, to use the correct term, co-operate.

Mr Maddocks did not think that such a scheme was workable.

Mr Rotherham: The question is whether by forcing co-operation as we can in war time we should get any practical results. If they do not accept the scheme then we must deal with the individual cases.

The Tribunal adjourned the cases to enable Mr Maddocks to consult his clients with a view to the adoption of a scheme of co-operation.[10]

In April 1916, an editorial in *TUJ* reported on the serious shortage of undertakers when it noted that, 'The authorities, in their zeal for recruiting, appear to be perfectly indifferent as to what become of the dead – and undertakers are becoming greatly alarmed as to conditions in the near future.'[11] William Spicer commented to a reporter that funerals were being delayed and, in some cases, businesses were being closed as all male members of the family had been called up. He continued:

> 'Funerals are having to the postponed every day for want of men to make coffins, and the other day a baby was taken to a London cemetery, but there was no grave ready to put it in. The funeral party had to put the body in the church and come back two days later for the interment.'

To which he added:

> 'It is not that we cannot get women for our work. Women could, of course, be used for inside work – the polishing of coffins, and so forth – but they cannot be employed for outside work ... It is all very well ... for chairmen of Tribunals to talk airily of being sewn up in sacks [a member of the tribunal in Whitstable had suggested that the dead be buried in sacks], but one knows that in the case of a funeral, sentiment is stronger than on any other occasion, and people will insist on having things done decently and in order.'[12]

Finally, Spicer noted that the war had prohibited the import of black horses from Flanders, and that in one case the motor hearse had '... come to the rescue.' As will be seen below, the motor hearse was having an increasing presence on funerals. Interestingly, in view of the need for horses at the Front, undertakers appeared to be unaffected by any shortage; there is only one brief mention of this issue in *TUJ* over the four years of conflict.[13]

For those firms appealing at tribunals, it was a matter of justifying the need to retain manpower and for chairmen to appreciate the labour-intensive nature of the work. Some were more willing to take this on board. However, as the war progressed and the need for soldiers increased, the degree of sympathy diminished.

There is evidence to show that women were engaged in funeral service. In October 1917, for example, *TUJ* reported that female bearers in Nottingham were provided with a 'smart uniform' and were said to be ' ... giving complete satisfaction to all concerned.'[14] Earlier the same year, the Edinburgh Town Council Parks Committee agreed to employ women, as they were at the City of London Cemetery.[15]

The delays to funerals in the Liverpool area in May 1916 were not helped by strikes by what was described as 'undertakers' men' – coffin makers and drivers – who demanded that Sunday funerals be stopped; the latter being a preferred day by the poor for interments. Employers rejected this claim, but the action of 200–300 workers led to the postponement of all services scheduled for Sunday 7 May. At a meeting held the following day, workers agreed to the compromise of returning as casual workers:

> *'Thus they do not break faith with their union, they do nothing to interfere with the negotiations which are still going on between them and their employers, and have made a concession in the interest of the health of the community.'[16]*

Sunday burials were stopped in Liverpool from June, and this decision seems to have been widely adopted. Only the previous year, a number of cemeteries, such as the City of London, had made the decision to prohibit Saturday afternoon funerals. NACS surveyed London cemeteries to test opinions and although met with a mixed response, they were supported by colleagues in the BUA.[17] Curiously, support for Sunday closing came from national newspapers.

In May 1916, the London Centre of the BUA had a meeting to discuss the treatment of undertakers before military tribunals. They passed a resolution, which was sent to Lord Kitchener and Andrew Bonar Law, that:

> *'... London Undertakers ... strongly urge the Government, for sanitary reasons, to refrain from further depleting the ranks of the profession by the obligation of military service. They also respectfully beg to warn the Government of the very grave danger which is threatening the public health through*

such depletion. The matter is of urgent national
importance.' [18]

One member, T W Muzzell, also presented a petition to be signed
and handed to local medical officers of health (MHO):

> 'We, the undersigned, Master Funeral Undertakers,
> respectfully solicit the assistance of Dr.........,
> Medical Officer of Health of the Metropolitan Borough
> or Urban District of ...., in bringing to the notice
> of the Government the necessity of a sufficient number
> of undertakers and their assistants remaining in the
> district to ensure burial of deceased persons before
> they become a menace to public health.
> I, ... Medical Officer of Health of the Metropolitan
> Borough or Urban District of........., declare my
> opinion that the funeral undertakers and their
> assistants at present in the above district are necessary
> to public health.
> Signed ....'

It is unclear whether this strategy was adopted by any MHO.

At the Centre's AGM, Mr Muzzell also informed members that
he had advertised for staff, only to receive one application by a man
with a wooden leg, which he thought of employing '... as a trade
curiosity (laughter).'[19] At the same meeting, Spicer reported that
progress was being made with canvassing MHOs and undertakers
to help get undertakers' employees on the list of exemptions.[20]
However, it was the attitude of tribunal chairmen that frustrated
those making appeals, as a *TUJ* editorial commented:

> 'What is the general, prevailing opinion of those who
> sit on those Tribunals? Contempt, indifference! There
> is nothing in it or about it worthy of the name of a
> trade or calling. That at least appears to be the opinion
> of the major part of those who sit in these seats of the
> mighty (or mite-y). Anybody can bury – knock a coffin
> together (sic), put a body in a sack, and shove it in a pit,
> which anyone can dig. Well, supposing it comes one
> day to municipalizing funerals, getting a bill through
> Parliament to that end, how would the Undertaker
> stand in front of such a contingency? He might have
> a little better chance than when the last bill touching

burial, undertaking and the rest was before the House
of Commons – when he was nowhere – actually and
pathetically nowhere!' [21]

Not all tribunals were, however, antagonistic towards undertakers. At the BUA conference that year, it was reported that those in Manchester made every effort to assist.[22]

The start of 1917 was marked by what was termed a 'strike' by Roman Catholic clergy in Liverpool, although this was actually a reduction in the times pubic funerals would be read each day at denominational cemeteries.[23] The main occurrence, however, was a deputation received by Neville Chamberlain, Director-General of National Service on Friday 12 January.[24] The MP for Bermondsey, Harold Glanville, facilitated the meeting before the 17 members of the BUA were introduced by the newly appointed BUA secretary, James Hurry. Edwin Dottridge explained the rational for the deputation:

> *'The crux of the whole matter is due to the fact that,*
> *unfortunately, our trade was omitted from the*
> *schedule of reserved occupations. This applies equally*
> *to the manufacturers throughout England...'*

He was not restrained in making sure the consequences were understood:

> *'I regret to say the local Tribunals throughout London*
> *and the country have been extremely unreasonable,*
> *and in many cases failed to realize that the nature*
> *of our business being that of national service, has*
> *been utterly ignored. The result is that a number*
> *of undertakers' shops have closed down, and great*
> *inconvenience has been caused to the trade.'*

In respect of cemeteries, one undertaker reported that he knew of '... a cemetery where the superintendent and whole of his staff had been called up, leaving the work in the hands of temporary employees.' He also noted the risk of graves collapsing after being excavated by inexperienced men.[25]

In his response to the deputation, Mr Chamberlain requested, ' ... definite facts of cases where bodies had been uncoffined for a longer time than usual, not owing to the neglect of relatives giving their order, but to the inability of the undertaker, through shortage of labour, to coffin the body in the usual time.' He also asked for information about delays at cemeteries. Hurry inserted a request in *TUJ* for this information.

But the delays to funerals continued.[26] In the same month, Bermondsey Borough Council conducted an inquiry into the problem before issuing a recommendation that the government, via the Local Government Board, prohibit the use of elaborate oak and elm coffins; they could be replaced by deal cases covered with cloth.[27]

TUJ continued to report on the ruling of tribunals, such as these from February, March and May 1917:

> 'The appeal of Thomas Samuel Horlock, 37, a Northfleet undertaker, passed for general service, who said he conducted 230 funerals last year, making all the coffins, was dismissed not to be called for one month.'[28]

> 'At Bermondsey Tribunal, W Uden & Sons, Ltd., undertakers, of Southwark Park Road, appealed for the exemption of Ernest Uden, secretary and manager. Mr Uden said in the early part of last year, when appealing for an employee, he warned the Tribunal of the dangers likely to arise through the lack of men employed in the trade, in that borough. They let all their single men go, with the result that they were now unable to cope with the work. He formed one of the deputation who waited on Mr Neville Chamberlain, who was impressed with the seriousness of the situation. The Mayor said he thought the Bermondsey Tribunal had been very lenient with the claims for exemption of men working in this trade. Mr Uden said he did not think so with regard to his firm. Their foreman coffin maker had to join up in a few days, and they would then be left with only one skilled man. It was only because they had an enormous stock of coffins in hand that they had been able to cope with the burials during the past months, and even then bodies had to wait four or five days. The military representative remarked that there were six brothers in this firm, and not one was serving. Mr Uden said that although the members of the firm were not serving, their sons were, one having lost a leg. Six months' exemption was granted.'[29]

'Mr Thomas Viger, undertaker, of London, a contractor with the War Office for the interment of soldiers who died in the greater part of the London area, complains that the House of Commons Appeal Tribunal has refused the exemption of his last able-bodied man.

"My contract with the War Office," he said in an interview, "covers interments from 82 military hospitals within a radius of ten miles of Charing Cross, and represents an average of 39 funerals a week. I have also contracts with the Canadian, Newfoundland, Australian, and New Zealand contingents, in addition to my civil obligations. My best workman, my clerk, and my son had previously joined up and now my motorman who also assists in making coffins, engraving and polishing the plates, and placing the dead in the coffins – has to join up on March 22nd. The manager of my branch establishment in Crawford Street is also before his Tribunal. I am now left with two feeble men over 60 years of age to carry on my interment work.

"Though it will be impossible to fulfil my contracts, the War Office has power to charge me with any additional cost they may be put to in interring the soldiers. Women cannot be expected to undertake the work, and I cannot get another man as a substitute anywhere." ' [30]

During the BUA conference at Brighton that year, Hurry revealed that he would be lobbying members of the House of Commons for the purpose of ensuring undertakers were on the list of exempted trades.[31] However, as can be seen from these reports, he continued to make representations:

'I mentioned my work of appearing on behalf of members either for themselves, their sons, or employees before the Tribunals. This was a very difficult piece of work, because as soon as it was known that I was acting for some firms, others felt that it would be better for them to relieve themselves of the help of solicitors who have been appearing for them and ask me to act. This meant travelling all over the country ... My appeal was always on the grounds that the work of a funeral director or his employee was work of national

importance, and in the majority of cases I was able to
satisfy the Tribunal to that effect.' [32]

The following shows Hurry's contribution:

'Mr Clements, an employee of Mr W S Bond [a well-
known west London undertaker], receiving month
final (sic). This is being appealed against and will be
heard before the Appeal Tribunal shortly. Mr H R
Sinclair appealed for his son and one coffin maker.
Three months was granted in each case. I hear the
Military are appealing against the decision. Mr H
Smith appealed at the Central Tribunal for his son
H C T Smith; although we did our best, he was only
granted one month. In the afternoon of the same day,
before the same Tribunal, Mr A J Evans appealed for
his brother, R Evans, and we got his case adjourned for
two months. Mr Cribb, of Canning Town, appealed at
the Appeal Tribunal for his second son, only just over
20, and classed general service. Again, a good fight
was put up and in the finish we gained the day and
obtained conditional exception.' [33]

In August 1918, the BUA finally received notification from
the Ministry of National Service that from the middle of that
month, the following categories would be included on the list of
certificate occupations:

Undertaking Trade:
Man wholly employed as coffin maker, as coachman or
chauffeur or as attendant.
Coffin making, Wholesale:
Foreman, Woodworking machinist.
Cemeteries, Public:
Man wholly employed as gravedigger.[34]

However, problems continued to occur, such as at cemeteries in
Manchester where conscientious objectors – 'conchies' – were
engaged to prepare graves, despite their inexperience.[35] Shortage
of manpower was compounded by the introduction on 22 July
of timber control: a permit was required to obtain supplies.[36]
The serious position was outlined in *The Times* along with the
suggestion of a substitute coffin made from papier-mâché.[37]

At the time, cremation held little favour as an alternative to burial: in 1915 only 1,410 cremations were carried out at the fourteen crematoria in operation. However, the shortage of gravediggers encouraged one journalist to commence a discussion on this topic towards the end of 1916:

> 'When reading this morning that the [Sheffield] Corporation was appealing for a gravedigger, and that older men could not be got to do the work, the claims of cremation came to my mind. Surely this would be a good time for people to put their unreasoning prejudices aside and adopt the modern method of disposing of the dead. Cremation is hygienic, economical of land and labour, and undoubtedly the coming universal method. We have a crematorium in working order. It is necessary, on the grounds of public health, that the present most wasteful and unsanitary method of interment should be abandoned.' [38]

Further support for cremations came from *The Daily Mail* and the *Evening News* at the start of January 1917.[39] The bold suggestion of 'compulsory cremation' was even made. The superintendent of Golders Green Crematorium was stated as claiming that two members of staff would be able to cremate sixteen bodies a day in two cremators: a fact disputed by John Robertson, superintendent at the City of London Cemetery and Crematorium.[40] Cremation came to the fore again when in June 1918, *TUJ* noted the irony of the London Cremation Company discontinuing its propaganda during the war years only to have the mantle taken up by correspondence in newspapers, such as those in the *Aberdeen Journal* and the *Evening Express*.[41] In addition to challenged attitudes towards burial, it would be the cost of cremation that would be a considerable deterrent.

A shocking issue that was first reported in the *TUJ* in January 1915, along with many other periodicals, was the burning of German corpses. After being stripped, dead bodies were then bundled into four with iron wire, before being transported to a facility north of Rheims where they were 'converted' into oil, fertiliser and pig fodder.[42] It was, however, later proven to be propaganda, with the basis being confusion over the mistranslation of the word '*kadaver*' that applied to both humans and animals.

FUNERALS DURING THE WAR

Despite the shortages and limitations brought about by the war, funerals continued. In addition to the domestic work, there was the burial of prisoners of war and also of British soldiers who died after returning from the Front. In many cases, they were given a military funeral as the images on page 75 show.[43] Paddington Council permitted free interment for British soldiers, whilst an area was set aside for the burial of Belgium soldiers at St Mary's Cemetery, Kensal Green.[44]

Whilst multiple deaths were occurring abroad, mass funerals at home were not unknown. The bombardment of Scarborough in December 1914 was an early tragedy necessitating the burial of a large number of casualties.[45] However, it was explosions at TNT factories that resulted in a number of tragedies. On 2 April 1916, 15 tons of TNT and 150 tons of ammonium nitrate ignited at the Explosive Loading Company factory near Faversham in Kent. The death toll reached 116 people.

Two similarly devastating explosions occurred the following year. On the evening of 19 January 1917, a fire ignited 50 tons of TNT at the Brunner/Mond munitions factories in Silvertown in East London. The blast was heard over 100 miles away and the fire visible for 30 thirty miles. It resulted in the deaths of 73 people. James Hurry was responsible for supervising one of the mass funerals that were held over a few days at the East London and West Ham Cemeteries.[46] The second was the detonation of five tons of TNT at the Hooley Hill Rubber and Chemical Factory in Ashton-Under-Lyne on 13 June. There, 46 people died, including the proprietor. The following Sunday, a public funeral was held on the steps of the town hall.

The largest death toll due to an explosion was in July 1918, when 134 people were killed at the National Filling Factory No 6 at Chilwell in Nottingham.[47]

As mentioned above, although no shift towards cremation of the indigenous population occurred during the war years, this mode of disposal was utilised in 1915 for 53 Hindu and Sikh British Army soldiers who died at hospitals in Brighton and Hove. A *ghat* was built at a remote site near Patcham on the South Downs. *TUJ* described the process:

> 'The burning is done on a funeral pyre of wood logs, in precisely the same manner and with the same ceremonies as those performed in India. The cremation

◄ Damage to a property in Scarborough following a raid by a German airship in December 1914

▲ ◄ In 1916, 105 people died in an explosion at Faversham, Kent. This photograph shows some of the 108 coffins buried in a mass grave at Faversham Cemetery. The committal service was conducted by the Archbishop of Canterbury. A white marble memorial covers the grave

◄ Some 13 horse-drawn hearses were assembled in front of Ashton Town Hall with 50 mourning carriages positioned behind, following the explosion at Ashton in June 1917. An estimated 250,000 were present for the funeral and lining the route of the procession to the cemetery. When the last mourning carriage had entered the cemetery, the gates were locked so that only families of the dead could be present for the burials

is conducted by a member of the same caste as that to which the dead man belonged.' [48]

In February 1921, the Prince of Wales dedicated a permanent memorial in the form of a Chattri, which that marks the location of the cremations.

REPORTS FROM THE FRONT

Throughout the war years, *TUJ* showed its support for those at the Front in a number of ways. A 'Roll of Honour' started in February 1915, and this monthly listing gave the name and rank of undertakers who had been called up. However, as time progressed, the inevitable obituaries appeared. Each Christmas, the London Centre of the BUA appealed through the pages of *TUJ* for funds to provide presents for members of the trade serving in His Majesty's Forces at home or abroad. The 1917 'Christmas Comforts Fund' raised £90, and parcels were sent – worth 17s 6d – each containing items selected by the ladies of the committee.

A further feature was the inclusion of accounts written by undertakers of their war experiences. All were subject to approval by the censor; most were from unspecified locations or the equally vague 'somewhere in Flanders'. With a few exceptions, descriptions were invariably positive and cheerful, almost to the point of falseness. The first to be published was, 'Christmas Truce for Burials':

'Christmas has come and gone, certainly the most extraordinary celebration of it any of us will ever experience. There were stacks of presents for officers and men, and no lack of comfortable hampers full of good things. In the yard of the farmhouse where my company was billeted there is a huge cauldron. In this no less than 125 lb. of pudding in tins was boiled at a time. The next day we returned to the trenches groaning under loads of comestibles and condiments destined to alleviate our lot on the morro. That night it froze hard, and Christmas Day dawned on an appropriately sparkling landscape. A truce had been arranged for the few hours of daylight for the burial of the dead on both sides, who had been lying out in the open since the fierce night fighting of a week earlier. When I got out I found a large crowd of officers and men, English and German, grouped around the bodies, which had

DEDICATION OF THE INDIAN CHATTRI BY H·R·H. THE PRINCE OF WALES ON THE DOWNS · FEB·1·1921

G.A.WILES BRIGHTON 5

▲ The dedication of the Chattri on the Sussex Downs by the Prince of Wales in February 1921. In 1915, 53 Sikh and Hindu soldiers were cremated on this site in 1915

◀ Britain's worst-ever rail accident occurred during WWI. Three trains were involved in the disaster that occurred on 22 May 1915: a troop train, a local train and the night express coming north from Euston. The troop train carried the Leith-based 7th Battalion Royal Scots Territorial Force destined for Liverpool. Signalmen at Quintinshill Junction gave clearance to the local train whilst also permitting the movement of a second troop train. The latter was compressed to half its normal length whilst also overturning onto the northbound tracks. The express from Euston then collided with the trains, causing an immense fire. In total, three officers, 29 non-commissioned officers and 182 soldiers were killed or burned to death. This image shows the coffins on their way to the cemetery at Leith

◀ 'Our Roll of Honour' published in *TUJ* in February 1915. Note the five members of the France family: Bert, Sydney, Henry, Charles and George

Our Roll of Honour.

Pte. Albert James Ames.
Of Messrs. W. A. Hurry & Sons, The Grove, Stratford.

Sergt. Harold Bacon, 6th Rifle Battalion King's Liverpool Regiment, stationed at Blackpool.
Harold Bacon is the only son (and child) of Mr. P. H. Bacon, of the firm of Ingall, Parsons, Clive & Co., Ltd , Liverpool. He joined when Antwerp was taken by the Prussians, saying "his King needed him more than J.P.C. & Co." Is 17½ years of age and was consequently not long in the business, but well known in Liverpool business circles.

Pte. William John Bailee, Royal Irish Fusiliers, Armagh Regt.
Coachman at Mr. R. R. Loudan's, Armagh.

Pte. William Bothwell, Royal Irish Fusiliers, Armagh Regt. Now at the Front.
Chauffeur at Mr. R. R. Loudan's, Armagh

Pte. Tom Broome, Mounted Sapper, Royal Engineers, West Riding, Yorkshire.
Son of Mr. Joseph Broome, Manchester. Pte Broome joined soon after the war commenced and expects to be called to the front soon. Stationed at present near Doncaster.

Pte. J. C. Broome, with the Public Schools Corps, at Leatherhead, Surrey.
Son of Mr. J. C. Broome, of 42, Downing Street, Manchester, President of Manchester Branch B.E.S.

Pte. T. Broome, Manchester Scottish, stationed at Edinburgh.
Son of Mr. Thomas Broome, Manchester.

Pte. Arthur Dyer, L.C.E. (Walthamstow), R.A.M.C., attached to the 2nd King Edward's Horse, stationed at Woodbridge, Suffolk.

Lance-Corporal William J. Dyer, National Guard.
Brother of the above.

Pte. Henry Edwin English, London Scottish.

Pte. Chas. Frederick English, 4th Cameron Highlanders.
Both members of the firm of W. English & Son, Funeral Furnishers and Monumental Masons, Bethnal Green Road, London, N.E.

Pte. Bert F. France, L.C.E., 1st Battalion London Scottish. Wounded at Ypres.

Pte. Sydney France, 1st Battalion London Scottish. Wounded at Ypres.
Both are sons of Mr. Albert France, Funeral Director, of 45, Lamb's Conduit Street, Holborn, W.C., and have been serving with the British Expeditionary Force since September last.

Pte. Henry France, R.A.M.C., now in France.

Pte. Charles France, King's Royal Rifles. Wounded at Zillebeer.
Brothers of the above Mr. Albert France.

Pte. George France, 5th Dragoon Guards, now with British Expeditionary Force.
Son of Mr. George France, Undertaker, Union St ,W.

Pte. John Garth, 1st Royal Dragoons, stationed at York.
Son of Mr. C. W. Garth, cabinet maker and undertaker.

Orderly Sergt. Thomas Garrity, R.M.C., serving at the Military Hospital, Fort Pitt, Chatham, with rank of Corporal.
Embalmer for Messrs. R. McDougall & Co., Liverpool. Till the War broke out was assistant hon. secretary of the St. John Ambulance Brigade, Liverpool Centre; was called up for duty in August.

Pte. Thomas Johnson, "A" Company, 5th Battalion Royal Inniskilling Fusiliers, Richmond Barracks, Dublin.
Son of R. S. Johnson, M.B.E.S , South Shields. Was one of the first in South Shields to answer Lord Kitchener's call.

Pte. E. C. Johnson, City of London Royal Fusiliers.

Pte. L. D. Johnson, Canadian Westmount Rifles.
Son of Mr. W. Johnson, Roman Road, Bow.

Pte. David Linton, Royal Irish Fusiliers, Armagh Regt.
Coachman at Mr. R. R. Loudan's, Armagh.

Pte. Loudan, Royal Irish Fusiliers, Armagh Regt.
Son of Mr. R. R. Loudan, of Armagh.

Lieut. E. Halford Mills, R.A.M.C., British Expeditionary Force, France.
Ex-President of British Embalmers' Society (London Branch). 30, Heath Street, Hampstead, N.W.

Corpl. H. E. Pierce, R.A.M.C., Chelsea Duke of York's School.

Corpl. S. W. Pierce, R.A.M.C., stationed at Egypt.

Sergt. H. S. Pierce, R.A.M.C., stationed at Bradfield, Norfolk.
The above are three sons of Mr. H. E. Pierce. Undertaker, 27, Featherstone Street, City Road, N.

Lieut. and Quartermaster O. C. Purnell, R.A.M.C., 2nd unit 1st Welsh Field Ambulance.
Of Cardiff. Member B.E.S. Council.

Pte. John Ritchie, Royal Irish Fusiliers, Armagh Regt.
Manager to Mr. R. R. Loudan, of Armagh.

Pte. Walter Scott, Leicestershire Yeomanry.
Son of Mr. F. Scott, Uppingham.

Pte. Walter S. Sherry, R.A.M.C. Duke of York School, Chelsea.
Son of Mr. H. A. Sherry, Treasurer, London Centre. B.U.A.

Pte. Dudley Sinclair, 6th Batt. Seaforth Highlanders.
Son of Mr. K. J. Sinclair, National President, B.U.A.

Corpl. J. H. Smith, Motor Cycle Despatch Rider, Royal Engineers.
Of Messrs. Higgins & Smith, Birkenhead. Died of injuries.

Pte. Frank W. Spicer, London Rifle Brigade.
Only son of Mr. W. G. Spicer, Secretary, B.U.A.

Pte. Bernard Stiles, Army Service Corps, Aldershot.
Son of Mr. C. Stiles, Manchester.

Gunner Samuel Thomas Sweeting, 15th Reserve Battery of the 6th London Brigade Royal Field Artillery, Territorial Forces.
Son of Samuel J. Sweeting, of Holland Street, Brixton, and previously engaged in the business with his father.

Colour-Sergt. W. E. Waugh, 5th King's Liverpool Regiment, at present at Blackpool.
Branch Manager to Messrs. R. McDougall & Co., County Road, Walton, Liverpool.

N.B.—Will all who can, send additions to "Our Roll of Honour" ? A post card, stating name, regiment and location, and any further particulars will oblige.

*already been gathered together and laid out in rows.
I went along those dreadful ranks and scanned the
faces, fearing at every step to recognize one I knew.
It was a ghastly sight. They lay stiffly in contorted
attitudes, dirty with frozen mud and powdered with
grime. The digging parties were already busy on the
two big common graves, but the ground was hard and
the work slow and laborious.'* [49]

The following year, an undertaker serving with the Royal Army
Medical Corps (RAMC) wrote to give his experiences of a night-
time burial:

*'We reached a little village just near the trenches,
which has been shelled into ruins by the enemy, and
here in a little house broken by shrapnel fire, we found
one or two wounded. Five men of the -- Regiment had
been killed during the day, and these we had to bury
in an estaminet [shabby café or bistro] garden. The
chaplain had come with us. During the burial service
he switched on his electric pocket lamp for a moment
or two. The instant he did so snipers began to fire at us.
The bullets whizzed overhead one after the other, all
quite unpleasantly near us. It was my first time under
fire by a graveside – a weird experience, but interesting
very. Two or three officers and about ten men stood
around the graves, and only one, as far as I could see,
moved a muscle during the firing. He bent his head
once almost involuntarily. One can quite understand
these chaps who have been out since the beginning of
the war, and going home on furlough soon, not wishing
to get sniped at the last moment, and I was surprised
at the coolness of everyone around that grave. The
chaplain told me afterwards that the officer said the
bullets had been very near to us.'* [50]

The four sons of George Comber, an undertaker from Redhill,
were in France during 1915 and extracts from their letters reveal
the destruction of towns by the Germans, the appalling weather
and the souvenirs acquired from the battlefields, for example, a
German helmet.[51]

As Chapter 1 noted, a key figure in the promotion of
embalming was an East-London-based undertaker, Arthur

▼ ▶ Private Dyer (standing in white smock) with the RAMC in 1914. He also sent sketches of the interior and exterior of his hut (*TUJ*)

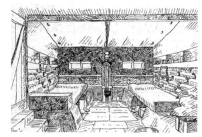

◀ Sapper George J Comber, photographed in 1915

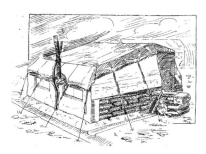

Dyer. He trained in London before completing further studies in the US; on his return he was active in the BES. Along with W Oliver Nodes, Dyer founded the London College of Embalming in 1913. However, by October 1914, he had joined the RAMC and was attached to the 2nd King Edward's Horse Regiment. In January the following year, Private Dyer wrote to *TUJ* to say that his knowledge of anatomy and first aid had been invaluable in assisting medical officers with tasks such as inoculations. In August 1915, he wrote of a subject close to home whilst giving an insight into the means of future identification of the dead:

> *'We have had several burials. They are very impressive, and necessarily very simple, but just the burial a soldier would wish. I have prepared nearly all the cases for burial, and in one instance dug the grave, practised ... with the Canadian contingent, is to place a rough cross with name, etc, upon it, and in addition a bottle is buried with the remains, in which particulars, i.e. name, number of regiment, date of death, etc, written on a piece of paper, are placed. Another bottle is placed near the surface with the same particulars. This, in my opinion, is the best that can be done at the time.'* [52]

A further contribution to *TUJ* was of sketches of the hut Dyer occupied in 1917.[53] He became a prisoner of war in March 1918, but was home by mid-December.

The two sons of Albert France of Holborn – Bert and Syd – served in the 1st Battalion London Scottish during 1914. Their letters home appeared in *TUJ* the following March. Although mentioning the losses, the focus was ' ... our exciting night of the 31st, including the heroic actions of the Company's doctor...' and the '... vivid recollections of burning houses, haystacks, trenches, hordes of Germans wounded and dying...'[54]

Finally, as the war progressed, *TUJ* published details of arrangements for soldiers' graves and the British Government pay for establishing and maintaining British military cemeteries abroad.[55]

THE IMPACT OF WAR ON FUNERALS

Whilst a number of writers have determined that WWI was the catalyst that had an impact on funeral customs, there is evidence to suggest that this change had already been ongoing since the late nineteenth century.[56] An editorial in *TUJ* of April 1914, largely cribbed from *The Times,* noted that, 'Funerals are daily becoming simpler' and that ' ... little by little the outward and "crying" signs of mourning are disappearing.'[57] *The Times* did note that the poor ' ... cling affectionately to old customs,' and patronisingly observed that they ' ... enjoy the self-importance of bereavement ... even the trip to the cemetery, is a treat.'[58]

The funerary historian, Julian Litten, summarises the overall attitude generated by the experience of war and its effect on mourning:

> '... there was a particular undercurrent of public opinion to contend with: the morality of staging a grandiose funeral when those who had died for King and country on foreign fields were unable to be repatriated. At such a time of great national suffering and sorrow, individual displays of funerary pomp and panoply did not sit comfortably on the conscience.'[59]

The sociologist Geoffrey Gorer expresses this even more succinctly: 'As the grief became more intense, and bereavement more widespread, the ostentation of mourning declined even further.'[60] This is confirmed by the costume historian, Lou Taylor, in respect of a '... breakdown in funeral and mourning etiquette.'[61] She adds:

> 'As the war continued, the survivors had somehow to face up to the loss of almost a whole generation

▲ Military funerals during WWI. A funeral procession led by the band of a light infantry regiment

▲ A Royal Artillery funeral

▲ A funeral taking place somewhere on the Western Front early in the First World War. At least two of the mourners are carrying the Glengarry cap, indicating that they belong to Scottish regiments. The coffin is resting on a hand bier

▲ A firing party at a burial in Stretford Cemetery, Manchester

▲ ▶ A funeral passing through the town of Dalbeattie, Scotland, and the scene at the graveside

of young men and the creation of a new army – this
one of widows and fatherless children ... Full ritual
mourning dress seems not to have been worn down to
the last detail. It was partly a question of morale, both
for troops on leave from the trenches and the public at
large remaining at home.' [62]

The historian Pat Jalland believes that the wearing of widows'
weeds '... would demoralise the nation.' [63] As some drapers had
links with the undertaking business – such as department
stores, including William Whiteley and Army & Navy – it is not
surprising that midway through 1918 the editor of *TUJ* wrote:

'We are strongly of the opinion that undertakers would
be wise to dissociate themselves from the mere drapery
part of funeral trappings and trimmings, we might say,
with the mere camouflage of sorrows and mourning;
for it is, in a large proportion of cases, nothing more
than that.' [64]

The second factor was the transition to motor hearses. Although
having a modest presence since the early years of the twentieth
century, the requisitioning of horses to the Front coupled with
the ban on plumes, helped to advance this change in ostentation.
In 1914, *The Daily News and Leader* commented that the motor
hearse was:

'... now an everyday feature of the modern funeral. It is
usurping the ancient prerogative of the horse; invading
the most cherished and sacred traditions ... With the
advent of the motor ... the tedium of long-distance
funerals is avoided ... A speed of 15 miles an hour is
regarded as the maximum rate of progress, but the
smooth gliding motor of the cars, the absence of beat
of hoof and rattle of wheels, and the only sounds an
occasional hoot on the hearse horn, do not suggest an
impression of unseemly hustle as would be created by
a horse-drawn funeral travelling at the same rate.' [65]

Presented with an opportunity to move towards motor vehicles, it
is not surprising that Dottridge Bros dispensed with advertising
their horse-drawn hearses on the front cover of *TUJ* and replaced
them with examples from their fleet. However, not all undertakers

▲ Coffins containing Belgium soldiers about to leave from the London Hospital in Whitechapel during the early years of the War

were supportive of the move towards motorised transport. At the 1915 BUA Conference, it was noted that Charles Porter from Liverpool realised that motor traction was bound to come:

> '... it was inevitable, but they [undertakers] need not hasten it or encourage it. And why? He did not want to appear selfish, but they had their interests to protect, and they knew that if motor traction was rapidly brought about, the horses, carriages, and harnesses, and all the paraphernalia, if stopped, became a very serious loss, involving thousands and thousands of pounds.' [66]

Writing in 1917, a cemetery superintendent summarised changes he had seen over the previous 25 years:

> '... from the old, ugly and heavy closed hearse, festooned with velvet, crowned with enormous black feathers ... with the coffin covered in black cloth, studded with black nails ... to the modern, light, open cars, the polished wooden coffins ... with the attending undertaker dressed in more rational garb.' [67]

▼ Dottridge Bros advertised this motor hearse after their horses had been requisitioned in 1914 (*TUJ* August 1914)

As mentioned above, the discontinuation of plumes was an example of this reduction in display. In 1917, the writer C E Lawrence noted:

> 'The war has greatly helped a tendency which was slowly gathering force. It is only necessary to hasten the certain end and so be done with the odious business of ordered ugliness and dark materialism of the conventional funeral.' [68]

The pro-cremationist Lawrence would have to wait some years before any significant move from burial would be recorded.[69] It was, however, the belief of some wartime tribunals that if staff were reduced, so too would ostentation: when responding to a request from C Ambler, an undertaker from Birmingham, for a request for '... enough men to bury our dead with decency and decorum,' the deputy commissioner for trade exemptions wrote:

> 'It is hoped, however, that undertakers will not overlook any means of economising labour and of employing workmen of a less skilled type than usual to relieve the situation. In other directions also there is room for economy of manpower in connection with funerals, but in respect of the polishing of coffins and in the general arrangement of funerals.' [70]

The chairman of the Law Society Tribunal, Sir John Paget, was even more forthright during an exchange with Hurry:

> 'If they [undertakers] cut down unnecessary extravagances on funerals they would go a long way to solve this labour difficulty. All these fellows in top hats and black tie walking along by the hearse might be working at the benches in the shops. You see them going along the streets with a lot of swagger. It makes me sick.' [71]

What the same unidentified newspaper failed to mention was a rejoinder to this comment:

> 'Alderman Hurry said that only two days ago a man not used to the work was engaged to help carry the coffin from the church to the hearse. He slipped and

lost his hold on the coffin, which nearly fell on top of the other bearers, who, by a back-breaking effort, managed to hold it.' [72]

Whilst pressure brought about by the war did go some way to encouraging the reform of funerals, the industry was anxious to continue its own agenda through the promotion of sanitary treatment of the body. Although the war had diminished the momentum fostered by BES and its pioneers, some educational classes were organised: the opening session in September 1916 of the London College of Embalming – held in Sheffield – was reported to have taken place against a background of air raid scares.[73] Further classes were conducted the following year by Alfred Morrison under the auspices of the Lancashire and Cheshire Divisional Council's educational programme.[74] The tireless Welsh pioneer, Morgan R Morgan, toured Britain with an embalming lecture tour during 1917.

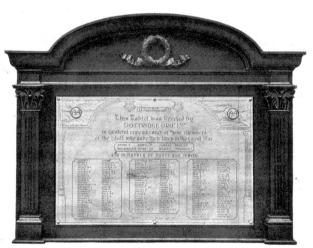

◄ Many staff from Dottridge Bros served during WWI, although only eight were killed. This memorial to those who lost their lives – and also those who survived – was erected at their City Road premises in 1921. It can now be seen at J H Kenyon, Westbourne Grove in Paddington

Messrs. Dottridge Brothers' War Memorial.

WE have much pleasure in reproducing, with Messrs. Dottridge Brothers' permission, a photograph of the Memorial which they have erected in their Head Office to perpetuate the memory both of those members of their staff who gave their lives in the Great War, and of those who served. In all 160 served, and out of such a large number, they were particularly fortunate in only losing eight, although a large number were wounded, or suffered from shattered health.

Mr. Roland Dottridge was very seriously wounded in Palestine whilst serving with the 2/16th London (Queen's Westminsters), but after two years in Hospital, and undergoing many operations, it is pleasing to know that he is restored to complete health.

Mr. Russell Dottridge had his health seriously affected for some time by shock and exposure following on his service in France with the H.A.C.

The Tablet is of brass, the names and inscription being engraved and waxed, and the pillars and scroll-work engraved only. In the corners, two bronze bosses bear the dates, 1914 and 1919. The mount is of mahogany, handsomely carved; the size of the Memorial is 4-ft. 9-in. by 4-ft. The Inscription reads as follows:—

This Tablet was Erected by
DOTTRIDGE BROS., LTD.,
In Grateful Remembrance of those Members
of the Staff who gave their lives in the Great War.

| BROWN, S. | DORRELL, F. | JAMES, A. E. | PRICE, F. E. |
| BULWINKLE, A. | GURNEY, E. E. | JESSOP, G. | TODHUNTER, J |

AND IN HONOUR OF THOSE WHO SERVED.

By receiving training and providing sanitary treatment of the body, undertakers gained the opportunity to enhance their status by distancing themselves from the uneducated traders of the past. Embalming was promoted as hygienic and in the interests of public health; the same approach would be taken by those advancing cremation.[75]

Whilst embalming occupied the minds of certain undertakers, the BUA continued with its annual conferences during the war years, despite postponement calls and very modest attendances. Nevertheless, strategic advances were achieved: at Blackpool in 1915 the amalgamation of BES and BUA was discussed while, as Chapter 2 has indicated, the association voted to seek trades union registration, which it achieved in December 1917.

By the time the war was drawing to a close, and having experienced a significant depletion of staff and then shortage of timber, undertakers would face one final challenge: the Spanish Flu.

1. Spicer WG (1943) '85 year old pioneer talks on early trade organisation' FSJ October pp267–268

2. 'Everyone's Aid Needed' (1916) TUJ March p67. See also 'BUA London Centre. 12th Annual General Meeting' (1916) TUJ June p152

3. 'United Kingdom Association of Cemetery Superintendents. Annual Congress' (1916) TUJ September pp255–256

4. Parsons B (2005) JH Kenyon: The First 125 Years Worthing: FSJ Communications p13

5. 'The Military Tribunals' (1916) TUJ April p87

6. Ibid

7. Ibid

8. 'The Military Tribunal' (1916) TUJ May p115

9. 'The Military Tribunal' (1916) TUJ July p192

10. 'The Military Tribunal' (1916) TUJ October p282

11. 'Shortage of Undertakers' (1916) TUJ April p88

12. Ibid

13. See Mitchell R (2011) 'War Horse' Britain at War Issue 60 pp65–71. See 'Horse Shortage in England' (1917) TUJ February p43

14. 'Personal and General' (1917) TUJ October p274

15. 'Women Gravediggers in Edinburgh' (1917) TUJ April p104. See Superintendents Day Book held at the City of London Cemetery 23 May 1916 p89 and minutes of the City of London Sanitary Committee Col/CC/PBC/01/02/020/16 23 May 1916

16. 'Undertakers' Assistants' Strike in Liverpool' (1916) TUJ May p117. See also 'Unburied Dead' Liverpool Echo 8 May 1916 p5 and 'Undertakers Man Return to Work as Casual Labourers' Liverpool Echo 9 May 1916 p6

17. 'Notes' (1915) TUJ August p235

18. 'British Undertakers' Ass'n. London Centre' (1916) *TUJ* June p152. See also 'Notes' (1916) *TUJ* May p115

19. 'The 12th Annual General Meeting' (1916) *TUJ* June p154

20. Ibid

21. 'Editorial' (1916) *TUJ* July p187

22. 'Thirteen Annual Convention of the British Undertakers' Association' (1916) *TUJ* August p224

23. 'Liverpool Burials. Roman Catholic Clergy Go on Strike' (1917) *TUJ* January p5

24. 'The Burial Crisis. Interview with the Director-General of National Service' (1917) *TUJ* January pp21–22. See also 'Funeral Delays' (1917) *TUJ* January p13

25. See 'Difficulties at South London Cemeteries' (1917) *TUJ* January p4

26. 'Held Up Burials' (1917) *TUJ* January p11

27. 'Bermondsey Burial Delays' (1917) *TUJ* February p48

28. 'The Military Tribunals'(1917) *TUJ* February p54

29. Ibid

30. 'The Military Tribunals' (1917) *TUJ* March p82

31. 'Fourteenth Annual Convention of the British Undertakers' Association' (1917) *TUJ* June p174

32. Hurry JR (1934) 'The National Secretary tells the History of the British Undertakers' Association' *BUA Monthly* September pp58–63

33. 'Before the Tribunals' (1917) *TUJ* May p133

34. 'Funeral Trades in the List of Certificates Occupations' (1918) *TUJ* August p199

35. 'The Unburied Dead. Undertakers at Their Wits' End' (1918) *TUJ* August p198

36. 'Timber Control'(1918) *TUJ* August p200

37. 'Timber for Coffins. Undertakers' Serious Position' *The Times* 17 August 1918 p3 and 'Papier-Mache Coffins. Suggested Substitutes for Elm' *The Times* 19 August 1918 p3

38. 'Shortage of Gravediggers' (1916) *TUJ* November p291

39. 'No One to Bury the Dead. Cremation for All' (1917) *TUJ* January pp25–26 (reprinted from the *Daily Mail* 3 January 1917 p4) and 'If I Could Choose' (1917) *TUJ* January p26l and 'Cremation as a Cure' (1917) *TUJ* February p52 (reprinted from the *Daily Mail* 12 January 1917 p4, and letter 15 January 1917 p4)

40. 'Some Questions Anent Cremation' (1917) *TUJ* February p52

41. 'The War and Cremation (1918) *TUJ* June pp161–162. See 'Letters to the Editor: Cremation' *The Aberdeen Daily Journal* 11 May 1918 p4, and 13 May 1918 p4,15 May 1918 p5 and 16 May 1918 p5

42. 'Concerning Burial: Germans Buried in Bundles' (1915) *TUJ* January p4; 'Burning German Corpses' (1916) *TUJ* April p110;, 'A Disgusting Way of Disposing of the Dead' (1917) *TUJ* May pp135–136; 'Editorial' (1917) *TUJ* June p142; 'Germans and the Dead' (1917) *TUJ* June p152. See also Ponsonby A (1980) 'The Corpse Factory' *The Journal for Historical Research* Vol 2 No 1 p121

43. 'Funeral of a German Prisoner' (1915) *TUJ* February p34

44. 'Notes' (1916) *TUJ* November p290

45. 'Scarboro' Funerals' (1915) *TUJ* January p4

46. 'The Great Explosion. Funeral of Dr Angel' (1917) *TUJ* February p33. See also Hill G and Bloch H (2003) *The Silvertown Explosion: London 1917* Stroud: Tempus Publishing. Nock OS (1973) *Historic Railway Disasters* 3rd Edition London: Arrow Books

47 Haslam MJ (1982) *The Chilwell Story* Nottingham: RAOC Corps

48. 'Funeral Pyres on the Downs' (1915) *TUJ* September p236. See also 'An Indian Funeral. A Burning-Ghat on Sussex Downs' (1915) *TUJ* November p314

49. 'Concerning Burial: Christmas Truce for Burials' (1915) *TUJ* January p4

50. 'A Side-Light on the War' (1915) *TUJ* February p42

51. 'Experiences of Our Roll of Honour Men' (1915) *TUJ* May pp127–128

52. A Letter from the Front. Corpl Arthur Dyer in the Trenches' (1915) *TUJ* August pp215-216;

53. 'Experiences of One of Our Roll of Honour Men' (1915) *TUJ* February p42. Arthur Dyer contributed on further occasions: 'A Friend from the Front' (1917) *TUJ* March p77; 'News from Second Lieut Arthur Dyer' (1918) *TUJ* July p189; 'A Letter from Germany' (1918) September *TUJ* p240; and 'Home from Germany' (1918) *TUJ* December p312

54. 'Experiences of Our Roll of Honour Men: Two London Scottish' (1915) *TUJ* March pp71–72. For further recollections see 'Experiences of Our Roll of Honour Men: Pte WN Docking' (1915) *TUJ* August pp216–217; 'Pte Murray' (1915) *TUJ* August p217; 'Letters from Lieut CB Waters' (1918) *TUJ* June pp158–159

55. 'British Soldiers' Graves' (1916) *TUJ* October p265; 'Prince of Wales on the Care of Soldiers' Graves' (1917) *TUJ* February p33; 'Soldiers Graves' (1918) *TUJ* March p54; 'The Soldiers' Graves' (1918) *TUJ* December pp305–306. See also Longworth P (1985) *The Unending Vigil: The History of the Commonwealth War Graves Commission* Barnsley: Leo Cooper

56. Cannadine D (1981) 'War and Death, Grief and Mourning in Modern Britain' in J Whaley ed *Mirrors of Mortality* London: Europa

57. 'Editorial: No more Mourning' (1914) *TUJ* April pp107–108. See 'Editorial: Signs of Mourning' *The Times* 14 March 1914

58. See also Cannadine D (1981) 'War and Death, Grief and Mourning in Modern Britain' in J Whaley *Mirrors of Mortality* London: Europa p193

59. Litten JWS (2002) *The English Way of Death: The Common Funeral Since 1450* London: Robert Hale p171

60. Gorer G (1965) *Death, Grief, and Mourning in Contemporary Britain* London: The Cresset Press p6

61. Taylor L (1983) *Mourning Dress: A Social and Costume History* London: George Allen & Unwin p266

62. Taylor (1983) p267. See also 'Mourning Suggestions' *The Lady* 29 March 1917 p339

63. Jalland P (1999) 'Victorian Death and its Decline: 1850–1918' in Jupp PC and Gittings (eds) *Death in England: An Illustrated History* Manchester: Manchester University Press p251

64. 'Editorial: Mourning Camouflage' (1918) *TUJ* May p127–128

65. 'Disappearance of the Horse in the Modern Funeral' (1914) *TUJ* January p9

66. 'Motor Funerals' (1915) *TUJ* July p187

67. Cochrane WA (1917) 'Some Funeral Customs' *TUJ* September p247–248

68. Lawrence CE (1917) 'The Abolition of Death' *The Fortnightly Review* Vol DCII 1 February pp326-331

69. The decline in mourning (along with the preference for cremation and donations in lieu of flowers) continued well into the interwar years. The trend was assessed using information from the obituary columns of *The Times* in 1930 and 1936. See letter 3 January 1930 and 30 March 1936 and also 'Analysis from "The Times" Funeral Wishes' (1936) *TUFDJ* April p112

70. National Archive NATS 1 1189 T. 1/770 26 January 1918

71. Unidentified and undated newspaper cutting (October 1917). NATS 1 1154. The wording differs to the version reproduced in *TUJ*

72. 'Notes' (1917) *TUJ* November p281

73. 'Embalming at Sheffield and Manchester. Opening of the LCE's Winter Session' (1916) *TUJ* October p272

74. 'Northern College of Embalming. Opening of the First Course' (1918) *TUJ* October p259

75. See Bourke J (1996) *Dismembering the Male: Men's Bodies, Britain and the Great War* London: Reaktion Books pp210–252.

CHAPTER 5
BURYING ENZA: COPING WITH THE SPANISH FLU PANDEMIC

First identified in May 1918 and named 'the Spanish flu' – Spain being the only country willing to admit there was a problem – the origins of the pandemic are uncertain, but have been possibly attributed to pig farms established two years earlier by troops in Northern France. Although flu has been a regular visitor to Britain for decades, the pandemic of 1918 occurred in three waves: June/July, then a recrudescence in October/November and finally in February 1919. The second wave claimed by far the largest number of victims: the worldwide number of deaths is estimated to have been between 50 and 100 million; in England it was put at 228,000. As one doctor noted, 'It came like a thief in the night and stole treasure.'[1] The flu was airborne, and spread through sneezing and coughs, which gave rise to a playground song that concluded:

> 'I had a little bird
> Its name was Enza
> I opened the window
> And in-flu-enza.'[2]

The virus almost exclusively attacked those in the age range of 25–45, with gender being roughly equal. Once a fulminating pneumonia had set in, death was fairly rapid.

THE SPANISH FLU IN LONDON
In London, the first reports of deaths as a result of a recrudescence of the flu were in October 1918. It prompted local health committees to close schools, while legislation was passed to require cinemas' patrons to vacate the building between performances to enable the air to be flushed. Handbills were distributed as precautionary measures.

The peak in the number of deaths was reached in the first week of November, just before the Armistice. It was not encountered in all areas of the capital, but the East End was particularly affected.

The Borough of West Ham was the worst hit, as indicated by the following figures:

WEEK ENDING/NUMBER OF DEATHS

OCT	OCT	OCT	OCT	NOV	NOV	NOV
46	71	113	201	379	370	285

(Source: Minutes of the West Ham Public Health Committee 20 November 1918 p50)

The dense living conditions in these areas were probably the main reason for the scale of contraction. In 1918, the population of West Ham was 294,523.

The pattern of mortality was similar in the west London area of Willesden, with the weeks ending 2nd and 9th November being the worst:

WEEK ENDING/NUMBER OF DEATHS

OCT 12	OCT 19	OCT 26	NOV 2	NOV 9	NOV 16	NOV 23	NOV 30	DEC 7	DEC 14	DEC 21	DEC 28
3	15	68	105	93	57	40	16	16	14	3	8

(Source: Willesden Urban District Council: *The 43rd Annual Report of the Medical Officer of Health for 1918*)

The same can also be said for the County Borough of Croydon:

WEEK ENDING/NUMBER OF DEATHS

OCT 5	OCT 12	OCT 19	OCT 26	NOV 2	NOV 9	NOV 16	NOV 23	NOV 30	DEC 7	DEC 14	DEC 21
1	5	20	46	110	95	55	38	31	19	13	4

(Source: County Borough of Croydon: *Annual Report of the Medical Officer of Health & Schools Medical Officer for 1918*)

In Tottenham, north London:

WEEK ENDING/NUMBER OF DEATHS

OCT 12	OCT 19	OCT 26	NOV 2	NOV 9	NOV 16	NOV 23	NOV 30	DEC 7	DEC 14	DEC 21	DEC 28
43	51	57	71	43	18	16	21	18	13	7	9

(Source: *Annual Report of the Tottenham Medical Officer of Health for 1918*)

In addition to the shortage of staff highlighted in the previous chapter, undertakers had to cope with all these additional deaths. In 1918, there were 730 members of the London Centre of the

British Undertakers' Association.[3] Virtually all were single-unit operations carrying out a small number of funerals, perhaps 50–70 a year. Some would have made their own coffins or purchased them finished from suppliers; the majority hired their transport resources from carriagemasters.

The increased number of funerals was revealed in a number of records held by undertakers:

WILLIAM DENYS

YEAR/MONTH	OCTOBER	NOVEMBER	DECEMBER
1916	17	19	31
1917	18	18	21
1918	34	36	26
1919	21	19	22
1920	23	31	27

In West Ham, the records of William Denys show that during October 1918, the firm carried out 34 funerals. This was just under double the number of funerals managed in the same month the previous year; in November it was exactly double, at 36. During October, the average age of the person buried was 35 years and the genders were roughly equal (M = 16 / F = 18). The business was about a quarter of a mile from West Ham Cemetery, and Denys's records indicate that from the total 96 funerals managed between October and December, 52 took place at the cemetery.

C E HITCHCOCK

YEAR/MONTH	OCTOBER	NOVEMBER	DECEMBER
1916	90	111	159
1917	101	106	112
1918	175	417	193
1919	120	96	123

A similar pattern was apparent from the records of C E Hitchcock of Barking Road. The majority of funerals were taken to the East London Cemetery about a third of a mile from their office. On 15 November, C E Hitchcock carried out 28 funerals, a huge number for a small business.

J Jeffries in East Ham was founded in 1902 and traded at the other end of Barking Road from C E Hitchcock. Their records show a busy October, but a quieter November; this point will be discussed below.

J JEFFRIES

YEAR/MONTH	OCTOBER	NOVEMBER	DECEMBER
1916	8	15	12
1917	5	4	10
1918	22	9	9
1919	7	8	11
1920	10	11	12

The Willesden firm of James Crook had a particularly busy November:[4]

JAMES CROOK

YEAR/MONTH	OCTOBER	NOVEMBER	DECEMBER
1917	25	29	39
1918	75	111	45
1919	45	34	39

While the funeral records provide an overview of the volume of work, so too do those in cemetery offices. The busiest days were during the weeks ending 9 and 16 November 1918.

Historically, the City of London Cemetery has always received burials from many parts of the East End of London. In 1918, there were 3,800 burials in the cemetery.[5] As the area was badly hit by the flu, the increase in monthly interment figures is not surprising.

CITY OF LONDON CEMETERY

YEAR/MONTH	OCTOBER	NOVEMBER	DECEMBER
1917	232	242	263
1918	367	807	371

The busiest day at the cemetery was 11 November 1918, when 67 interments were recorded. On four others days that month, there were in excess of 50 burials.

Between 1 October and the end of December, 957 interments were undertaken at West Ham Cemetery; the busiest day was Thursday 14 November, when 30 burials took place.

WEST HAM CEMETERY

YEAR/MONTH	OCTOBER	NOVEMBER	DECEMBER
1918	221	512	224

It was, however, the privately owned East London Cemetery in Barking that undertook more burials than any other cemetery in this area of London. The cemetery received funerals Monday to Saturday and, during November 1918, the average was 46 interments per day. The busiest week began on Monday 11 November.

EAST LONDON CEMETERY

MON 11	TUES 12	WEDS 13	THURS 14	FRI 15	SAT 16
56	74	56	58	84	31

▼ The City of London Cemetery at Ilford: 807 burials were carried out in November 1918

The busiest day was Friday 15 November, when 84 burials were recorded, with Hitchcock's bringing nearly 30 funerals to the cemetery.

Over a four-year period, it is possible to see the impact of November 1918 at the East London Cemetery.:

EAST LONDON CEMETERY

YEAR AND YEARLY TOTAL/MONTH	OCTOBER	NOVEMBER	DECEMBER
1916 (3276)	256	238	350
1917 (3331)	234	248	229
1918 (4551)	340	1208	426
1919 (2841)	245	214	249

The number of deaths in the Croydon area during November 1918 can be seen from the table below, which shows interments in the area's two cemeteries.

CROYDON QUEEN'S ROAD CEMETERY

YEAR/MONTH	OCTOBER	NOVEMBER	DECEMBER
1918	119	234	73

CROYDON MITCHAM ROAD

YEAR/MONTH	OCTOBER	NOVEMBER	DECEMBER
1918	110	221	158

(Source: The Croydon Advertiser)

Conversely, elsewhere, October was busier that November, as seen in this three-year analysis of burials at Tottenham Cemetery.

TOTTENHAM CEMETERY

YEAR/MONTH	OCTOBER	NOVEMBER	DECEMBER
1916	68	101	129
1917	88	92	113
1918	321	253	168

(Source: Minutes of the Tottenham and Wood Green Burial Board)

HOW DID FUNERAL DIRECTORS AND CEMETERIES COPE WITH THIS VOLUME OF WORK?

In many respects, it was fortuitous that the national secretary of the British Undertakers' Association, James Hurry, was an undertaker in West Ham. He was prominent in local government having been councillor, alderman and mayor of West Ham, in addition to being the third president of the BUA.[6] He negotiated the release of around 100 soldiers to help with shortages encountered by colleagues in funeral businesses, and cemeteries, by meeting with the National Service officials and the London Command.

Hurry also arranged for a one-page pro forma to be printed in *TUJ* that owners of funeral firms requiring manpower could complete and return to his office. However, by the time the edition appeared in December 1918, the worst of the second wave of flu had passed.[7] Nevertheless, he did assist in getting help from the military in London, Manchester and elsewhere.[8]

As Chapter 4 has already identified, James Hurry had already had experience of negotiating with civil servants and the director-

▸ *TUJ* included in their December 1918 edition the BUA's letter of request for additional staff. However, by the time it was published, the flu had subsided *(TUJ)*

▾ The BUA Secretary, James Hurry (1866–1945). Photo taken in 1909 *(TUJ)*

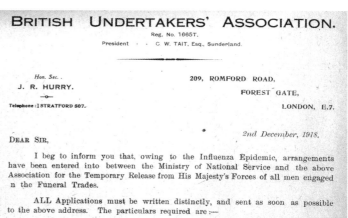

general of national service, Neville Chamberlain.[9] There were also other sources of help for undertakers. The *Stratford Express* noted that arrangements had been made for the local council's carpenter to assist local undertakers in the making of coffins.[10] In Manchester, an aircraft manufacturing firm agreed to produce coffins and turned out 40 a day.[11]

The second issue was reports of the shortage of wood for the construction of coffins, a situation heightened by the 1918 timber control order. More than one local paper reported that undertakers had difficulties getting a sufficient quantity of timber.[12] Another source states that London was even 'running out of coffins', but this is difficult to substantiate. One source said that the shortage of wood was largely due to government requirements for ammunition boxes and huts.

At the time, London was very well served by coffin manufacturers: Dottridge Bros had their own canal-side warehouse at Hoxton in East London, while Ingall, Parsons, Clive & Co's coffin-constructing factory was at Wealdstone near Harrow.[13] In addition, there were suppliers such as Henry Smith. Both Dottridge Bros and Ingall's would have had a large store of wood that was held for seasoning until winter, when it would have been used during the busy months. If wood was not seasoned or of poor quality, it would simply have been covered with fabric. If nameplates were not available, details would have been painted or written on the lid of the coffin.

The third point concerns claims of undertakers refusing work: one firm in Croydon had 32 orders in hand and allegedly declined eight others.[14] Another in Edmonton had 63 orders and had to refuse a dozen more.[15] Whilst there is evidence to support this claim, it was more a case of 'Come back later' rather than 'We can't help at all'.

In the records of J Jeffries, a hiatus is discernable. They show a funeral carried out for a male dying on 29 October 1918, but then the next funeral entry is for a person dying on 7 November. Anecdotal evidence suggests that the business was so busy at this stage that a notice was put up in the window of the premises to say that no further funeral orders were being taken. If this was the case, Jeffries had a space of eight clear days to carry out the funerals remaining from October and then cope with the funerals of those dying after 7 November.

The fourth point is the issue of the delay between death and burial. At the time, it was customary for the coffin to remain at home until the day of the funeral. The only exception was if the body was removed to the public mortuary for a coroner's post-mortem.[16] Decomposition was inevitable and the closed coffin would remain in the room where the family lived. Custody of the body tended not to be within the remit of the funeral director; it was the interwar years before this changed. However, it would appear that even where a mortuary was provided by the local authority, it was not used. The Metropolitan Borough of Poplar recorded that their mortuary accommodated 336 bodies during 1918; of these, 286 were removed on the instructions of the coroner and only four at the request of friends of the deceased.[17]

Funeral records reveal the delay between death and burial. On 15 November 1918, C E Hitchcock undertook 28 funerals: the average interval of whole days (excluding day of death and the day of burial) was 9.1 days, with the shortest 7 days and the longest 12 days. This contrasts to the first five workings days of October 1918 (prior to the onset of the Spanish flu), when the average delay was 5.3 days.

Five points can be made concerning cemeteries. The first concerns volume. Although cemeteries normally worked a five and a half day week, it would appear that through necessity many interments were carried out on a Saturday afternoon. Few, however, were prepared to operate on a Sunday. East London Cemetery adhered to their policy, whereas the City of London did make an exception, but only with single burials being recorded on three consecutive Sundays during November. The reluctance to prepare graves on this day can be attributable to the necessity to give grave-digging staff at least one whole day off each week; the willingness of undertakers to book funerals on a Sunday; the availability of clergy who would be engaged in leading worship (although cemeteries did have chaplains); and adhering to the general tradition of not working on 'the Lord's Day'. Furthermore, as Chapter 4 noted, the BUA and NACS had pressed for Sunday closing. There is evidence, however, to show that staff at the Jewish Cemetery in Willesden worked one Sabbath to prepare graves for Sunday interments.[18]

The Borough of Willesden adopted a unique strategy to cope with the abnormal conditions arising from the epidemic, by making arrangements with the neighbouring Hampstead Cemetery, which was not so severely affected by the flu, to take some of the common interments. Willesden Council paid the difference between their common interment fee (19s) and the charges made by Hampstead (£2 2s).[19] In a few cases at Tottenham, the burial service was read, but the interment held over until the grave had been prepared.

The second point is that to an extent, cemeteries would have been used to high volumes of work, particularly at certain times of the year such as during the winter. For example, research indicates that on 1 January 1900, 64 burials took place in the City of London Cemetery, a figure not dissimilar to the daily totals experienced during October and November 1918.[20] The relentless volume of burials over approximately the three-week period in 1918 was unprecedented, but it did prove that cemeteries could cope under pressure.

There is no evidence to suggest that burials took place in communal graves: 'trench burials' or 'collective graves' as they have been termed. As Chapter 4 illustrated, this did take place in the case of multiple deaths from single events, such as after the Faversham explosion earlier in 1918.

Thirdly, although cemeteries were dealing with a high number of interments, many would have been in common (public) graves with multiple occupancy, and others in re-opened graves: a 20ft public grave may well have accommodated eight coffins. Not all interments would have necessitated the digging of a new grave. Furthermore, there was plenty of new grave space.

The fourth point concerns the sheer physical labour required to continuously prepare graves over a sustained period of time: all would have been hand-prepared. An insight into how staff coped is provided by the superintendent and registrar of the City of London Cemetery and Crematorium, John Robertson. He reported to the City's Sanitary Committee:

> 'About the third week of October the pressure of work became so severe that ... I telephoned to you asking for assistance to help with grave digging, and advised you to communicate with the military authorities .... the Eastern Command immediately sent a working part of twenty men, most of them used on land work, and consequently of great assistance.
> During the five weeks ending 21st instant, a total of

▶ John Duncan Robertson, Superintendent of the City of London Cemetery and Crematorium 1913-1936 and founder, first member and first president of National Association of Cemetery Superintendents

836 interments took place, as against the 284 in the corresponding part of last year, showing an increase of 552 interments in the five weeks...

On the 11ᵗʰ instant, 65 interments took place, and on the 13ᵗʰ instant, I had orders in hand for 139 interments...

During most of the time I have had to keep the office staff at work each evening to cope with increased registrations and other clerical work, and paid them overtime ... I have had to keep the best gravediggers at work every Saturday afternoon and Sunday.

I would like to point out to you that during this time nearly all other work in the cemetery has been at a standstill.[21]

All cemeteries would have employed both gravediggers and gardeners, and it is likely that the latter would have been seconded to grave-digging operations during the Spanish flu crisis. Like the City of London Cemetery, a number of cemeteries were helped by military personnel, while others were assisted by council staff. Tottenham Cemetery recruited the services of two men from the nearby military hospital, two from the Wood Green District Council and two from the Tottenham District Council. They were paid their usual wage and 1s 6d extra per day.[22] At the privately owned Abney Park at Stoke Newington, the local borough council provided a number of employees to help with grave preparation. Many of the grave-digging staff would have lived locally, so the problem of getting to work would have been minimal, particularly if the flu had interrupted bus or tram systems.

The fifth point is in respect of a religious service prior to interment. It is likely that many funerals took place at public reading times. This involved a number of hearses arriving at the cemetery chapel and the mourners (often only the immediate family) taking their seat whilst the burial service was read. The coffins would not be brought into the chapel and, at the conclusion of the service, the hearses would go to the prepared graves and wait for the minister to go from grave to grave to conduct the committal ceremony.[23] This would explain how one clergyman managed to read twenty services in one day at East London Cemetery.

The last issue is cremation. Despite London having three crematoria, the number of cremations was insignificant. In 1918, there were only 70 cremations at the City of London and only 1,795 in total at the 13 crematoria in operation in Great Britain.

This table shows the yearly number of cremations at the City of London Crematorium.[24]

1914	1915	1916	1918	1918	1919	1920
42	45	43	68	70	58	57

Clearly the public could not be persuaded to break with the tradition of burial. An undertaker writing in *TUJ* said that cremation would not help during the flu as crematoria could only cope with six cremations a day. He said that '...one cemetery will dispose of more corpses in a day than a crematorium.'[25] It seemed pointless challenging this attitude when such long distances had to be travelled to crematoria, and cremation was no cheaper than burial. In Sheffield, the issue of a free cremation was raised.[26] It would be a further 40 years before a significant move towards cremation was recorded.[27]

1. Quoted in 'The Influenza Pandemic' (1971) *Purnell's History of the First World War* Part 107 Vol 7 No 11 p2978

2. Quoted in Honigsbaum M (2009) *Living with Enza: The forgotten story of Britain and the great flu pandemic of 1918* London: Macmillan p77

3. Figure from the British Undertakers' Association *Year Book and Diary 1919*

4. '150 Years of Funeral Directing: James Crook Ltd Celebrates Anniversary' (1951) *FSJ* May pp283–284

5. Minutes of the Corporation of London Sanitary Committee November 1918 and January 1919 Col/CC/PBC/01/01/22 and 23

6. 'His Worship the Mayor of West Ham: Alderman JR Hurry' (1911) *TUJ* November pp305–6. See also 'Interview with Councillor RJ Hurry. Social Reformer and Undertaker' (1909) *TUJ* January pp13–15.

7. See also JR Hurry (1934) 'The National Secretary Tells the History of the British Undertakers' Association' *BUA Monthly* September pp58–63

8. 'The Armistice' (1918) *TUJ* November p289; 'The Epidemic and the Trade' (1918) *TUJ* November p291; and 'The Epidemic. The BUA and the Situation' (1918) *TUJ* December pp297–298

9. 'The Burial Crisis' (1917) *TUJ* January pp21–22

10. *The Stratford Express* 27 October 1918

11. 'Notes' (1918) *TUJ* December p293

12. *The Willesden Chronicle* 1 November 1918

13. Parsons B (2009) 'Unknown Undertaking: The History of Dottridge Bros: Wholesale Manufacturers to the Funeral Trade' *Archive: The Quarterly Journal for British Industrial and Transport History* September No 63 pp29–41

14. *Croydon Times* 6 November 1918

15. *The Tottenham and Edmonton Weekly Herald* 1 November 1918

16. See Fisher P (2009) 'Houses for the Dead: The Provision of Mortuaries in London, 1843–1889 *The London Journal* Vol 34 No 1 March pp1–15

17. Medical Officer of Health Report for the Metropolitan Borough of Poplar (1918) p20

18. *The Willesden Chronicle* 1 November 1918

19. Willesden Borough Council minutes, 26 November 1918 p393

20. Information obtained from microfilmed registers held at the Guildhall Library

21. Minutes of the Corporation of London Sanitary Committee November 1918 and January 1919 Col/CC/PBC/01/01/23. John Duncan Robertson was superintendent from 1913–1936 and first president of the National Association of Cemetery Superintendents (now ICCM)

22. Minutes of the Tottenham and Wood Green Burial Board 21 October 1918

23. West J (1988) *Jack West: Funeral Director, 60 Years with Funerals* Ilfracombe: AH Stockwell pp112–113

24. Minutes of the Corporation of London Sanitary Committee November 1918 and January 1919 Col/CC/PBC/01/01/22 and 23

25. 'The Epidemic and Cremation' (1919) *TUJ* January p25

26. 'Cremation at Public Expense' (1918) *TUJ* November p282

27. See Jupp P (2005) *From Dust to Ashes: Cremation and The British Way of Death* Basingstoke: Palgrave MacMillan.

CHAPTER 6
UNDERTAKEN WITH DETERMINATION: THE MILLER CASE

If ever an example existed of perseverance in the face of adversity it would be that of William Miller. Determined to establish himself in the funeral business after returning from the Great War, he was prevented from doing so by members of the British Undertakers' Association. Reaching court in December 1920, the judge found that 'a grave wrong' had taken place. Duly compensated, Miller progressed his business, which today continues to thrive.

Drawing from court reports, along with original documentation and funeral records, this chapter traces the extraordinary events of the long-forgotten Miller case.

HOW IT ALL STARTED

As Chapter 2 noted, the BUA was registered as a trade union on 28 December 1917; with this status came greater power to regulate the activities of its membership. At the time, most undertakers did not possess their own funerary transport and hired horse-drawn vehicles from carriagemasters. Those operating in the capital were members of the London Funeral Carriage Proprietors' Section of the BUA, and were only permitted to supply, at agreed rates, undertakers who were members of the association. This effectively meant that all undertakers had to be in membership with the association. The BUA also laid down a minimum scale of charges for funerals. For example, in March 1918, the cost of an elm coffin for an adult together with a hearse, pair of horses and bearers was £4 10s. The penalty for BUA members charging less than the agreed rate was a fine and, ultimately, expulsion.[1]

It was an allegation of conspiracy to prevent Miller from becoming a member of the BUA that led to the case in which James Hurry and Henry Repuke had to defend their actions. As described in the previous chapter, James Hurry was a Stratford-based undertaker, who had been involved in the affairs of the BUA and its London Centre since its founding. He served as national president in 1909 and, after being the London area secretary, was appointed as the full-time national secretary immediately prior to the Miller case reaching court; he occupied this position until retirement in 1934. Founded in the late 1880s, Henry M Repuke's business was located at 333 Upper Street, Islington: he also had three other branches in the vicinity. Commencing on 2 December 1920, the case was heard in the King's Bench Division before Mr

◄ William Miller photographed during WWI (courtesy of Peter and Andrew Miller)

▼ The BUA secretary, James Hurry (1866–1945), photographed in the early 1930s (*BUA Monthly*)

Justice Darling (1849–1936). Called to the Bar in 1874, he took Silk in 1885 and was known for his wit, which was evident during the trial.[2]

Acting for Miller was Mr J B Melville, who started by reviewing the background to the case when proceedings opened on 2 December 1920. Prior to 1908, Melville began, Miller was in the army but had left as a reservist. Between 1908 and 1912 he was the business manager for his father-in-law, Henry Merrett, who owned a funeral business in Bethnal Green. Miller left to start on his own at 48 Clarence Terrace in Lower Clapton but, after a year, was anxious to find alternative premises. Until 1914, he assisted other undertakers, at which point he took employment with Henry Repuke. Miller then rejoined the army but whilst on leave in November 1918 assisted Repuke, who was under great pressure from the number of funerals caused by the Spanish flu pandemic.

Following demobilisation, Miller informed Repuke he wanted to start his own business. Repuke's manager, Tom Potter, had left the company at the end of February 1919. Consequently, Repuke was very anxious to retain Miller. At news of Miller's plans, Repuke became very angry; he threatened to ensure that Miller was prevented from opening in the Islington area.

Miller knew that undertakers had to be members of the BUA and, on 7 March 1919, he wrote to James Hurry about the application procedure. In April he made formal application and enclosed the 1 guinea subscription. Shortly after, Miller entered into a yearly tenancy of premises at 95 Essex Road, where he intended to establish his undertaking business. A few days later, he wrote to the secretary about his application, but received no reply.

On 5 May, the London Centre's committee recommended Miller for election subject to him trading in Clapton, but not Essex Road.

As Miller did not hear anything further from Hurry, he consulted the weekly publication *John Bull*. At the time, it ran a column of letters asking searching questions aimed at business organisations and government ministers. Presumably Miller hoped that his cause would be taken up. The manager of the *John Bull* enquiry department wrote a letter to Hurry to say that the position was extremely unsatisfactory owing to the fact that no definite reason could be assigned for the delay to Miller's application of membership of the BUA. In the meantime, the Association had received opposition to Miller opening in Clapton from a local competitor, Mr Moss. The matter was adjourned until 14 June. However, on 2 June, Miller attended a BUA meeting where he revealed he had taken premises in Essex Road. Hurry said to him:

> *'It is impossible for us to elect you if you want to open a new business. Unless you are elected a member, it will be impossible for you to carry on business, because nobody will supply you.'*

must not supply the carriage. Merrett replied that he had already entered into a contract with Miller, to which Hurry said that if he did he would be reported and fined; non-payment would result in expulsion from the Association. (On a previous occasion, Merrett had been fined for allegedly supplying something under the agreed price; because at first he did not pay the fine, he had been boycotted until he did.) Merrett regrettably informed Miller that he could not carry out the contract.

Miller then contacted Mr Seaward, an undertaker in Tottenham, only to find that Hurry had told him that Miller was not a member of the Association. The carriagemaster and wholesale funeral supplier Dottridge Bros said they would be willing to assist, provided that Hurry was agreeable. Dottridge rang Hurry and, after asking if he could supply Miller, commented: 'I think it is a shame to prevent a man from earning his own living.' Dottridge did not supply Miller. It is unclear who carried out the funeral, although it is in the Miller records.

A further opportunity to sort out the matter was during a committee meeting held on 14 November, when Hurry revealed that Norton (the undertaker in Essex Road) was not going to start trading. William Miller was asked if he intended to join Repuke's former manager, Tom Potter, in business, which he denied. He was recommended for election and on 10 December attended a members' meeting to state his case. However, Repuke declared that Miller and Tom Potter intended to form a partnership. As they would take away his business, he asked for the Association's protection. Once again, membership was denied without Miller being given an opportunity of speaking in support of his application. After issuing a writ, he obtained an interim injunction where Hurry and Repuke gave an undertaking not to interfere with his business (this undertaking lasted until the start of the trial).

Miller was cross-examined by Mr Holman Gregory KC, who represented Messrs Hurry and Repuke:

> *Was it not by arrangement with John Bull that you got the order for the child's funeral in order to test the rules of the Association?* - No.
> *Did you talk it over with them?* - No.
> *You were willing to sign the form of application for membership of the Association, showing that you accepted the conditions of membership?* - Yes.
> *So if other people sign it, too, don't you think they*

*ought to observe the rules and conditions? - Yes, they
ought to.*

*Don't you think it was the duty of Mr Hurry, as
Secretary of the London Centre of the Association, to
see that the rules were complied with? - Yes.*

The case continued the next day, with Mr Holman Gregory
declaring there was no evidence of conspiracy. He acknowledged
that Repuke was anxious that Miller should not be allowed to
set up in competition and also to prevent membership of the
Association, but there was no evidence that Repuke and Hurry
were acting in collusion to this end and with malicious intent.
Pointing out that trade-union cases were always difficult and
recognising that litigation was expensive, Mr Justice Darling
suggested that the parties come to an out-of-court arrangement.
The judge also rejected Gregory's contention that there was no
evidence that could be put before a jury. However, although
neither the Association nor the London Centre were parties to
the litigation, any such arrangement would need to be considered
by the 900 members of the Centre, and a meeting for this purpose
could not be convened immediately.

Just before the jury was discharged and the court adjourned,
another quip came from the judge:

> *'...then it will necessitate calling a meeting of the
> members. I hope it will be done as soon as possible,
> or else, having regard to my age, they will have to
> bury me first. That might be a very appropriate end to
> the litigation.'*

A special meeting was held on 15 December, when 84 members
voted for Miller's application for membership to be approved, but
not if he remained in Essex Road.

The case was resumed on 17 December with Mr Holman
Gregory disclosing the decision of the meeting held two days
prior. It was at this point that the key reason for not approving
membership was given:

> *'...there is only a limited number of available funerals,
> [and] allowing too many persons to trade in that
> neighbourhood as members of the Society would not
> be right.'*

In his opening address, counsel for the defence acknowledged that Repuke did not think it was fair that Miller could open near him and that, if he could do so legitimately, he would stop him setting up as he believed insufficient work existed for another undertaker. He did ask two or three people to vote against Miller's admission but said that there was no harm in this action. James Hurry was described as a man of 'exceptional ability' who had made a significant contribution to the Association both during its early years and when it became a trade union. Hurry believed he had a responsibility to see that rules and sanctions were enforced. Repuke gave evidence that there was never any arrangement with Hurry that they should work together to keep Miller out of the association.

When questioned, Hurry said the Association was formed for the purpose of improving the status and knowledge of undertakers, and they sought to make every undertaker a sanitarian – an embalmer. Mr Justice Darling then enquired why Miller should not become a sanitarian. Hurry's response was that there was nothing to stop him from being one, but added that due to the number of undertakers in the area, it might not be worthwhile receiving the training. Mr Justice Darling asked:

> 'Why not allow the natural law of supply and demand to operate?'

Hurry's response was:

> 'Well, the funeral trade is different to any other.'

Hurry denied any desire to injure Miller. He did not know Repuke had asked people to vote against Miller's election, and denied encouraging Repuke to do this. Furthermore, he had no knowledge that Merrett had agreed to supply a carriage for the child's funeral and denied instructing him not to honour this arrangement. When questioned about how he responded when confronted with such a situation, Hurry responded:

> 'I am given to understand so-and-so has an order for a funeral; kindly note he is not a member of the Association.'

The final person to be cross-examined was Albert J Cottridge, an embalming teacher and member of the BUA. He acknowledged that Miller's application for membership had been dealt with in the ordinary way. When asked by Mr Melville for the reason why Miller was not elected, he confirmed that the basis for rejection was the existence of sufficient members of the trade engaged in that area.

THE JUDGEMENT: 'A GRAVE WRONG'

Mr Justice Darling commenced his judgement by briefly summarising the facts surrounding the case before focusing on the evidence of Merrett, who had proposed Miller's application. The judge identified the action taken by James Hurry to convince him that a conspiracy had been committed:

> 'I [Merrett] agreed to supply a carriage and pair of horses for the funeral. Later in the day, Hurry rang up on the telephone and said I was not to supply the plaintiff under any circumstances, as he (Hurry) was stopping his supplies all round, and I should be fined and boycotted if I did supply him. I told the plaintiff I could not carry out the agreement. I had already been fined ten guineas by the Association, and was boycotted until I paid the fine.'

Also noted was Hurry's similar remark to Dottridge. Consequently, the charges of conspiracy to prevent Miller from becoming a member of the Association were proved. Hurry and Repuke were '... in concert trying to prevent the plaintiff from becoming a member of the Association.'

The judge's remaining remarks had more than a critical tone. He believed that trade unions acted as though they were practically put above the law. Whilst recognising that the BUA was formed with '... some beneficent objects ... it [is] perfectly plain its real object was to keep up the prices which should be paid for funerals, and have as few people engaged in the business as possible, and therefore it was to be made a closed business as far as could be. If any unemployed people wanted to work, they would not be able to get it in the undertaking business.'

Miller's army service also influenced the judge:

> 'It seemed a peculiarly shocking thing that a man who ... had done his best in the war and suffered from it should, on coming back, be met with such a fence like this, set up to prevent him getting back into the

position in which he could keep himself ... If ever there was a man who deserved some encouragement when he came back to earn his own living, the plaintiff was that man.'

Finally, as William Miller estimated that loss of profit was £10 per week, compensation of £150 was awarded, with Hurry and Repuke also responsible for the costs. The judge granted William Miller an injunction to prevent continuance of the acts complained of. Counsel for the defendants applied for a stay of execution, but this was refused. The case was summarised in *The Times* with the concluding sentence containing one of Mr Justice Darling's infamous puns: 'A grave wrong had been done.'[3]

THE AFTERMATH OF THE TRIAL AND THE GROWTH OF WILLIAM MILLER'S BUSINESS

The decision not to allow William Miller to open on Essex Road was based on the fact that there were already many undertakers in the area. Research supports the BUA's case as, according to the Post Office Directory for 1919, there were 22 undertakers listed in the London N1 postcode district: the area catchment from which William Miller would attract funeral instructions. In the same year, there were 1,134 deaths in Islington. Dividing the number of deaths by undertaker premises shows that this would give each firm an average of 50 funerals. However, the thrust of the Miller case was conspiracy to prevent membership of the Association, not the issue of competition.

An editorial published in *TUJ* the month after the trial contained a guarded comment:

> *'Regarding the Miller judgment, it is being said that it is so unsatisfactory that it might be worth the BUA's while to appeal against it. We should doubt the wisdom of such a step were it entertained. There are unsatisfactory points about the judgment, but – let us remember, " 'Tis better to suffer the ills we have than fly to others that we know not of.' " Our point is ... we know now, by bitter experience, that though we are allowed to make stringent laws, we must not attempt to enforce them by too stringent measures.'*

TUJ also acknowledged, perhaps inaccurately, that there were no winners:

'We are sorry to find Mr Miller still out in the cold, if such be the fact. We are sorry for the defendants, for the Secretary of the BUA, especially, if through his [James Hurry's] zeal for the cause which he has so much at heart he was unwittingly led, as Mr Justice Darling assumed he was, to overstep the line of prudence and so lay himself open to the charge of conspiracy.'

TUJ also stated that there was not much to regret about the case, only the costs. In this point they were correct. Although the precise figure cannot be ascertained, the Association was forced to appeal for subscriptions from its membership to clear the estimated debt of £2,000. Discussion about this matter took place at that year's BUA conference. However, two years later, over £360 remained outstanding.

Ultimately, William Miller was successful in his venture. His company's funeral numbers gradually increased over the first years of trading: 15 in 1920; 62 in 1921; 55 in 1922 and 89 in 1923. A sad fact apparent from the registers is that many of the funerals were for children; in 1920, just under half were for those aged five and under; the following year they represented one third of the funerals. This corresponds with the number of child deaths in Islington: of a total of 4,640 deaths in the borough of Islington in 1919, 525 were for children aged less than one year old. This was also the lowest proportion for some many years.[4]

The funeral registers also reveal much about the manner in which funerals were managed. At the time the firm was founded, the majority of people died at home and the body was kept there in the interval between death and the funeral. However, the funeral records indicate that some deaths occurred at the Islington Infirmary on St John's Road and also in the Metropolitan Hospital. After receiving funeral instructions, William Miller or one of his members of staff would obtain a measurement of the body, return to Essex Road to construct a coffin made of elm (or possible oak) and then deliver it to the house, where the deceased would be encoffined. It's likely that a local woman would prepare the body, although there is no record of this in the funeral registers. For children, a small box, probably covered in white fabric, would be provided.

The majority of funerals carried out in the company's early years were burials at the Borough of Islington's Finchley Cemetery; a small proportion took place at the privately run cemeteries serving the north and east London areas, such as

Manor Park, Chingford Mount, Tottenham Park, Woodgrange Park and East London. In the early years, the burials taking place the furthest distance from the office were at Forest Hill in south London. At the time, cremation would not have been a common occurrence: in 1920, only 0.34 per cent of deaths in the UK were followed by cremation and there were only 14 crematoria, with three – Golders Green, the City of London and West Norwood – in the London area. Between 1920 and 1923, William Miller only arranged one cremation. Compared to burial, it was certainly an expensive option: his charges for the coffin and transport amounted to £16 16s, while the cremation fee alone was £6 16s 6d.

William Miller continued to hire carriages until such time as he acquired his own horses and hearses. However, this mode of transport was comparatively short lived as motor vehicles were gradually coming into use at funerals. A document in possession of the Miller family shows that their last funeral horses were sold at the Elephant & Castle horse auction in April 1937.

Henry Repuke died in February 1933 and James Hurry in March 1949. The BUA became the National Association of Funeral Directors and was no longer registered as a trade union. The number of funeral directors in the Islington area gradually declined in the second half of the twentieth century: Henry Repuke's business closed in the late 1970s. However, the family owned firm of W G Miller continues to trade from 95 Essex Road.

1. All extracts have been taken from the case reports in *TUJ*: 'Undertakers' Action over Alleged Intimidation. Milller v Hurry and Repuke' (1920) *TUJ* December pp373–375. See also 'Undertakers' Action over Alleged Intimidation' (1921) *TUJ* January pp5–7 and 'Special Meeting of the London Centre' (1921) *TUJ* January p8 and 'Editorial: The Miller Case and After' (1921) *TUJ* February pp59–60

2. See 'Obituary. Lord Darling: Wit and honour on the bench' *The Times* 30 May 1936

3. 'Law Report, Dec 17. Summary of cases' *The Times* 18 December 1920

4. See *Medical Office of Health Report for 1919* Borough of Islington.

CHAPTER 7
FROM CRADLE TO GRAVE: THE CO-OPERATIVE FUNERAL SERVICE

From their emergence in the seventeenth century to the latter part of the nineteenth century, the undertaking industry has been dominated by the sole trader. However, this trend was challenged in the 1870s when department stores started to offer undertaking services. Diversification was appropriate as stores such as William Whiteley, the Army and Navy and Harrods would be equipped to provide mourning wear to those arranging funerals. However, it should be noted that these organisation frequently sub-contracted their funeral work to a local undertaker or carriagemaster. Nevertheless, the entry of the department store into the funeral sector was significant, as it represented the first example of the separation of ownership from control.

The roots of industrial co-operation can be traced back to 1844 when the 28 members of the Rochdale Equitable Pioneers Society started to trade and share profits. They adopted a philosophy based on open and voluntary membership, democratic control (one member, one vote), a dividend, education, and the provision of quality goods and services.

There is limited evidence to show that a small number of the several thousand Co-operative societies in existence in the second half of the nineteenth century offered an undertaking service. Founded in 1861, the Lincoln Society was listed as 'joiners and undertakers' in the Co-operative Directory of 1893.[1] By 1900, the Hull Society was 'cabinet makers and undertakers', while the New Civil Service Co-operative Society provided a 'monumental masonry' service from 1887. In 1905, societies in Beldington, Blaydon and Alloa were also listed.[2] However, it would not be until the second decade of the twentieth century before funeral furnishing became a core service. By this time, the Co-operative movement was involved in coal mining, grocery, meat, dairy, pharmacy, footwear, photography, brushmaking, dentistry, ironwork, bee keeping, convalescent homes, hosiery, furniture, watch-making, publishing, brewing, printing, banking and insurance. This last service provides the clue to the movement's entry into funeral service. The Co-operative Wholesale Society (CWS) started to sell insurance in 1867 and offered life assurance 11 years later. However, it was not until 1913 when it became a wholesale function – the Co-operative Insurance Society (CIS) – that this service increased in importance. In 1917, the CIS absorbed the Planet Friendly Assurance Collecting Society and, with a house-to-house collection system in place, rapid progress was achieved. One service the CIS offered was the death benefit policy; it was a sure way of preventing a pauper funeral.[3] The amount paid out at death depended on the level of purchases from Co-operative outlets. For example, in 1935, if a husband died and a wife made a claim, then the payment was calculated on 4s per £1 purchased, to a maximum value of £40 – considerably more than the average cost of a funeral.[4] This posed the question of why the CIS was paying out money at death to members who would then use the services of a private undertaker to supply a funeral. Funeral furnishing was an entirely logical extension to the range of services: societies could literally supply members from cradle to the grave.

Unlike other undertaking firms, the barriers to entry for the Co-operatives societies were very low for the following reasons. First, members of the life assurance scheme who obtained the death benefit policy would arrange a funeral with the Co-operative, thereby generating a guaranteed income for the society. Secondly, in addition to those who had a death benefit policy, the Co-operatives had a loyal clientele who were likely to use another services provided by the Co-operative, especially if further dividends were obtainable. Thirdly, it is likely that the family would also spend money purchasing mourning wear or food for post-funeral refreshments at the Co-operative stores. The supply of funerals also encouraged diversification into the monumental masonry business, and also wreaths.[5] Fourthly, many societies had premises that could be adapted to accommodate a small funeral office. 'Behind-the-scenes' activities could also be serviced, including stabling or a garage for animate power or motor hearses, and a joinery shop that could be used to construct coffins. Furthermore, where a Co-operative hall existed, it could be used for the post-funeral refreshments. Fifthly, economies of scope could be achieved as staff could be drawn from other departments, such as deliveries or the joinery department, to assist on funerals, while limousines could be hired out for weddings. Sixthly, Co-operative departments already in existence could be utilised for supplies, such as timber for coffin construction – at one point, the CWS motor-body department designed and made a motor hearse, while the CWS timber works in Norfolk supplied wood for coffins (the fabric for coffin interiors could be had from Co-operative textile mills). Seventhly, a loan could be obtained from the Co-operative bank to establish a funeral service. Finally,

▲ A horse-drawn hearse outside a branch of the Birmingham Industrial Co-operative Society Ltd (Courtesy of Midlands Co-operative Archive)

societies with only a small membership could federate with other local Co-operatives to create a funeral service.

It was a combination of these factors that led many societies to establish a funeral department in the post-WWI years. The Nelson Society was reportedly carrying out half the funerals in the Lancashire town by 1919.[6] The following year, the society in Pendleton acquired an established business and then rebranded it as a Co-operative enterprise.[7,8] Eccles, Willington and Barnsley were all in operation by mid-1920, followed by York and Luton in 1921, then Sunderland and Newcastle two years later.[9] Ipswich started in 1925. Some smaller societies federated to supply services such as laundry and simply extended this to funerals: Ashton, Beswick, Denton, Droylesden and Failsworth are examples that commenced in 1926 and traded under the name of Manchester Co-operative Funeral Undertakers Ltd.[10] Bradford had opened three months previously when they released the following statement in November 1926:

> 'An annual death roll of our members ... represents
> more than ten per week. This solemn fact reminds us
> that our Funeral Furnishing Dept. is not getting its

▼ ▲ ▶ The Birmingham Co-operative Society opened a funeral service in 1937 and operated from purpose-built premises on the corner of Holt Street and Ashted Row. The large site comprised offices, reception rooms, living accommodation, coffin showroom, garage, coffin workshop and store, mortuary and chapel of rest. The garage housed two Daimler straight-nosed hearses and five limousines. The chapel was dedicated in November 1936 by the former Bishop of Persia. Demolished in the early 1990s, the glass in the chapel of rest commemorating deceased employees of the society has been preserved and awaits reinstatement *(TUJ)*

fair share of work. This department has been open three months, and has done fairly well, but not in proportion to the claims we have had for the Collective Life Assurance. If we got only half of these claims it would be a welcome improvement on the past ...'[11]

In 1925 there was support for a funeral service in Glasgow.[12] The first in London appeared in March 1929, when the Royal Arsenal Co-operative Service (RACS) launched its funeral service (see below). North of the Thames, the London Co-operative Society started in 1931.[13] In 1934, the Leicester Society and the CWS made a move into funerals, followed by Birmingham[14] a year later, and Walsall and District in 1936.[15] One interesting development is that a small number of societies acted as agents for another society or for a private undertaker.[16]

The historians of the Co-operative movement, Carr-Saunders et al, reported that by 1935 there were 122 societies engaged in undertaking.[17] Clearly, funeral service was an important development for the Co-operative societies as by this time the membership in Britain was more than six and a half million;[18] by 1939, it was estimated that at least 90 societies had federated to provide a funeral-furnishing service.[19] Three years later, it was estimated that collectively the societies managed 9 per cent of the 450,000 funerals in England, while by 1946 this had increased to 14 per cent.[20] Nor was WWII a deterrent to progress, as the South Suburban Co-operative Society funeral service was established at Purley in 1942.[21] By 1956, there were 379 societies offering a funeral service.[22]

HOSTILITY TO THE CO-OPERATIVE SOCIETIES

The Co-operative's diversification into funerals immediately caused friction with established undertakers, who regarded this development as a threat to their livelihood. There were many firms in existence – for example, in 1927 there were around 780 firms within a 20-mile radius of Charing Cross[23] alone – and the majority were members of the BUA. It was not long before *TUJ* was filled with correspondence; many contributed using a pseudonym. The action by carriagemasters in one town and the response by a new society were as follows:

'We read in the Northern Echo that owing to the refusal recently of livery stable keepers to supply carriages and horses to convey a coffin which was made by the

Sunderland Co-op Society, it has been decided to establish a complete funeral-furnishing department. The decision has come to be representative of practically the whole of the Co-op Societies in the Sunderland district, and a committee has been appointed to work at the details of a scheme.[24]

Refusing to supply societies with funeral transport was how the BUA dealt with Co-operatives' entry to the market. This decision was taken at a national level when, at the 1919 BUA conference it was resolved that:

'... no Co-operative Society, not now carrying on the business to-day, be accepted as members of the association. It would then be left with the Centres to deal with those who were in now.'[25]

Certainly, some centres did embrace societies. Correspondence in the trade press started in 1926 and a lively exchange followed in spasms over the next decade. In May 1926, the wisdom of the boycott extended to the Bradford Society was questioned; funerals were a competitive market and:

'... the undertaker who is already in the trade is in the position of any other tradesman with whom the Co-op enters into competition. He will survive if he is fit to survive.'[26]

When in the late 1920s the RACS announced that they were to establish a funeral service in Woolwich (see below), the London centre of the BFWA held a meeting to discuss the impact of this venture. Concerned about terms and conditions for coffin makers and drivers, the association did not believe the RACS was a threat and made contact to request the employment of men already engaged in the trade and not to recruit new staff.[27] One correspondent thought prospective clients would have confidence in a 'store' rather than in a person,[28] while another believed that:

'... while the Co-operative Societies could not at once produce the human qualities that are needed, they could bring into the funeral trade some organising ability which is urgently needed.'[29]

'A Northerner' admitted that funerals were too expensive as, in proportion to the number of funerals in some towns, there were too many carriagemasters and undertakers trying to sustain a living. The correspondent believed that societies would introduce 'more economical methods'. Indeed, an article in the Co-operative periodical *The Producer* noted that, 'One result of their entry into the undertaking business was a tremendous reduction in price so far as the general public were concerned.'[30] Additionally, if the societies were boycotted by the BUA, the managers would form their own association to further their knowledge of sanitary matters and funeral management.

During a debate at the 1929 BUA conference that took place under the heading of ' "Co-ops". An Increasing Danger', it was revealed that societies were exempt from paying income tax. The matter had been taken up by the National Chamber of Trade, but to no avail, and this situation would only be changed by the Finance Act 1933.[31] Meanwhile, William H Crook of London noted that the Co-operative Societies were now:

> '... building up immense resources; they were not only taking over businesses but were putting up immense structures which meant high finance and if they were not checked, the time might come when their activities would make a serious inroad on the available sources of taxation.'[32]

'A Northerner' returned to the pages of *TUJ* in September 1930, stating that the boycott should be abandoned as the BUA was powerless to prevent the societies offering funerals. He believed that the businesses that had withstood competition were the ones built on personality and the confidence that a family had in an undertaker:

> 'Personal attention, coupled with efficiency and a desire to please has laid wonderfully strong foundations in the past. It remains to be seen whether new competitors offering new inducements will affec old firms who are doing their work well.'[33]

In January 1931, financial support of an 'Anti-Co-oper Movement' was sought from undertakers. Its president, C Keys, said that:

> 'The entrance of the Co-ops into the realm of burials has not been accomplished without some indication of mass production and modernisation such as is evidenced by the scale of charges, trade lists and the common knowledge of the deceased person's after-death value, according to co-operative insurance schedules.'[34]

It was said that undertakers had no protection and should not be complacent, as had other commercial sectors that had suffered a loss of business to the Co-operative societies. Furthermore, success of the societies:

> '... spells R-U-I-N to all private enterprise and individualism and they can destroy the Constitution and Empire ... Their driving force is politics – insidious politics which have their roots in Marxian Socialism and a misguided conception of economics.'[35]

However, the *BUA Monthly* expressed caution by encouraging members to keep their money in their pockets '... and wait for a substantial plan of campaign ...' as the Societies' campaign organiser '... knows just as much about the funeral directing profession as he puts into written or printed words and that is nothing.'[36] Correspondents soon restated the well-rehearsed remarks about the effect of the Co-operatives on their businesses, the futility of a boycott and the lack of reaction by the BUA as conference debates had been held in secret.[37] 'A Lancastrian' soon changed the direction of the argument, however, when he commented on the consequences of the Co-operative on the established trade:

> 'The general effect in our town has been that the other undertakers have smartened up. They realise that they have a strong competitor and have not been so foolish as to adopt boycott. The result has been that while there is competition, there is harmony of working.'[38]

He continued by adding:

> 'The Society is not doing a large business ... They have not the personal factor. The old firms have. Those which are well managed and give intelligent service have scarcely suffered. Those who have lost business to

the Co-operative Society are the low-grade men.'[39]

In January 1933, a *TUJ* editorial heralded a flurry of letters after it said that businesses would survive '... by offering a different and more personal service ... Business, in the long run, will be found to flow, not necessarily to the biggest, but to the best.'[40] An 'Observer' wrote to say that many BUA centres were working with the societies and that some were loyal members of the centre.[41] The BUA's response was largely defensive:

> *'It will be well to the advantage of Members throughout the country to be wide awake as to the ramifications of the Co-operative Societies in their areas because there is no doubt that they have got their eye on the funeral business and before you know where you are, you will find business filched from you by their progressive methods.'[42]*

Whilst revealing a north/south divide between BUA members – those in London not having such competition from the societies as other colleagues – the correspondence also noted the Co-op's strong appeal to women as '... in the time of a serious domestic crisis, they are only too likely to fly to a place where they are well known.' This was particularly the case when a maximum death benefit of £66 13s 4d was being paid to members.[43] 'GPA' wrote to say that the opening of Co-operative funeral departments was 'inevitable' on account of the money paid out following a death. However, he also noted that:

> *'... they had started ... in the right direction ... the best rolling stock, offered attractive wages, and introduced good conditions, superannuations funds and clothing schemes for the men.'[44]*

Consequently, it attracted the best men who knew the shortcomings of the private enterprise. This position was confirmed when it was said that existing funeral businesses should learn from the Co-ops:

> *'In the face of serious competition, wise traders try to do better – to give something different, to give quality, and, above all, personal attention. They must "do better or go out".'[45]*

With the march of the Co-operative unabated by mid-1935, the *TUJ* editor called for a close working relationship between both 'sides' of the trade, and for all undertakers to operate ethically.[46] Attention was also drawn to the fact that an Association of Co-operative Society Managers had recently been formed.[47] However, with issues arising from WWII occupying the NAFD, it was not until 1945 that the question arose of the Co-operative societies being admitted to membership of the association. At a meeting in September, the usual areas of contention were aired, including the questionable one of a society being unable to give the same service as an individual funeral director, and the Co-operative providing 'free funerals' under the death allowance scheme. It was decided that the societies in London should not be admitted into membership.[48]

One correspondent wrote to *TUJ* pointing out that the ideals of the Rochdale pioneers had been lost and that the societies threatened the livelihood of the independent trader:

> 'Our profession is being prostituted by many of these societies, it being no uncommon sight to see a notice displayed in butchers, bakers, grocery departments, etc: "Funeral orders taken", and one of our members has reported that on one occasion in his area the Co-op milkman was the person who called for the measurement and instructions.'[49]

Concern was also expressed that Co-ops were purchasing businesses previously run by private traders, and figures were published showing the pattern.

In October 1945, two unidentified Co-op managers broke silence by writing to the *FSJ* defending their conditions of employment, the preference for employing BIE qualified embalmers, and reminding fellow undertakers that the Co-operative funeral service was there to stay:

> 'When the lads come home they will thank us for our stand against ridiculous prices, loungers waiting at street corners for some poor soul to draw the blinds and then run along to their pet undertaker for a ghoulish commission.'[50]

It was, finally, in 1948 that the Co-operatives were permitted to apply for membership of the London association.[51]

CO-OPERATIVE MANAGERS UNITE

Although an association for Co-operative managers had been established in 1935, it made little progress. Post-WWII, the Co-operative Funeral Service Managers' Association (CFSMA) found impetus and the first conference of the National Co-operative Funeral Directors' Association was held in October 1945.[52] Attracting around 100 delegates to Blackpool, the first president of the association, J M Lucas, general manager of CWS funeral service at Manchester, drew attention to the fact that in 1934, only 87 societies had a funeral service; in 1944, it was 320 who collectively looked after 80,000 funerals.

The first item on the agenda was education, and a report indicated that a possible approach would be in conjunction with the Co-operative Union and the CWS. It was noted that many members of the CFSMA were also members of the BIE, and that dialogue should be held to explore the possibility of the Co-op's syllabus being accepted by the institute. A national wage scale was also unanimously agreed.[53]

The following year, delegates returned to Blackpool where further details of the educational programme were given, including managerial and embalming courses.[54] At the meeting of the executive of the association in July 1947, the future of the training scheme was the key item for discussion, including the possibility of education being run by the Co-operative Union. By this time, a number of embalming classes were already being run under the auspices of the CFSMA. It would also appear that the NAFD national president, along with members of the LAFD, were anxious to 'appease' Co-operative societies to get their support for matters of national importance in recognition of the Co-op's ability to '... get the ear of the Government'.[55]

At the 1948 conference, it was noted that embalming was ardently advocated by the association. However, the managerial course had not been progressed to the extent that the executive had hoped.[56] It was at this meeting that the president, F H Wilkinson, sent a message of greetings to members under the banner 'Sincerity in Service', with a laurel wreath bearing that legend.[57]

By 1949, the proposed CFSMA's educational scheme was a three-year City and Guilds approved course with agreement that the CWS would give a grant to applicants to study for the BIE Diploma. By this time, some 50 students were undergoing training, such as at the weekend school in Embalming and Mortuary Technique held in Newcastle in July 1949.[58]

THE ROYAL ARSENAL CO-OPERATIVE SOCIETY

The RACS was founded in November 1868 as the Royal Arsenal Supply Association.[59] The decision to establish a funeral service was announced in September 1928 when *TUJ* reproduced a statement that appeared in *The Times*:

> 'The Royal Arsenal Co-operative Society, of Woolwich, announces that in the near future it proposes to institute a department to provide every service appertaining to funeral furnishing. The management committee say that they regard a funeral-furnishing section as a natural complement to the Society's death benefit scheme. It is surely inconsistent, the committee add, for the Society to be paying away about £14,000 a year in death benefits and to leave private enterprise to undertake the funeral arrangements.[60]

In the RACS' monthly *Comradeship and the Wheatsheaf* newsletter, progress of their latest service was charted. In January 1929, Frederick Henry Wilkinson was appointed to organise this new department and open the first branch at Plumstead. The language used was important: 'He will work on modern lines, and will not be hampered by obsolete equipment, but will

▶ 'Each for all and all for each.'
The former RACS headquarters
on Powis Street in Woolwich

▲ Frederick Henry Wilkinson, the RACS funeral manager

introduce all the better ideas culled from a wide experience of the most modern tendencies in this branch of work. Daimler funeral vehicles were purchased whilst the branch would have a mortuary chapel.' [61]

Described as 'dynamic, militant and aggressive', Wilkinson was the RACS funeral-furnishing department's manager from 1929 until he retired in 1958.[62] After distinguished war service, he became a funeral director in Eastbourne, before being appointed to the RACS. He was a founder of CFSMA and also its fifth president. As an active embalmer, he encouraged all his managers to qualify. He was also one of the first funeral directors to lecture on aspects such as service, psychology and ethics.[63]

The Plumstead branch opened in February 1929; by May it was said to be '... flourishing in South London in no uncertain manner ...' [64] The society then opened a chapel of rest at Woolwich, which became the headquarters of the funeral-furnishing department. By June 1929, letters of appreciation were being received at Powis Street, including one from E C Hendry, a former member of the RACS Education Committee:

▲ The RACS chapel of repose at Woolwich. The site on which the chapel stands was part of the old naval dockyard. During the reign of George IV, the dockyard was reconstructed by French prisoners of war and it is believed that the chapel was used by the resident staff as a place of worship. In 1927, the RACS negotiated the purchase of part of the original Royal Dockyard and the chapel entered into the society's possession. The chapel was renovated for use as a 'chapel of repose' when the funeral service was established in 1929. The building around the chapel was enlarged in 1962 to incorporate a memorial showroom, masonry workshop, funeral depot and garage. The altar backcloth and frontal came from the now-demolished Holy Trinity Church, Beresford Street in Woolwich

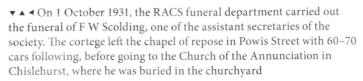

▼ ▲ ◄ On 1 October 1931, the RACS funeral department carried out the funeral of F W Scolding, one of the assistant secretaries of the society. The cortege left the chapel of repose in Powis Street with 60–70 cars following, before going to the Church of the Annunciation in Chislehurst, where he was buried in the churchyard

'... we expected it would be a fair service, and we were pleasantly astonished at finding it so first class. We cannot express what our feelings were to ... ... find such an absence of sordidness, the usual type of man and carriage.[65]

By the end of the second year of trading, the department reported a 40 per cent increase in revenue. A branch was opened at Greenwich, then Earlsfield.[66]

The first branch at Plumstead was opened in March 1929; a second at Woolwich followed two years later. The service then expanded into south-west London with a purpose-built office, chapel of rest and garage at Earlsfield. At an opening ceremony conducted in October 1933 by the Bishop of Kingston, the bishop commented:

'I rejoice we are doing something ... which will contribute to the reverent care of the departed in this area. For 25 years I was a parish priest, and during two-thirds of that period I worked amongst the very *poorest of the people. Again and again, I was filled with the greatest sadness at the conditions which existed when death visited the homes. It will be a great benefit if the dead can be taken from their homes to such a chapel of repose as we are dedicating today.'*

In August 1935, *TUJ* published a profile of RACS. The success of the department was attributed to the leadership of its manager, Frederick Wilkinson. It was noted that '... considerable care has been exercised in the selection of experienced men ... Every opportunity is afforded the young man desirous of becoming a first-class funeral director.' It was noted that five of the managers possessed the BIE diploma. Staff were paid union rates and membership of a union was essential. They also received a half-yearly bonus, regular annual holidays with full pay, and

▸ The RACS branch at Earlsfield. Designed by the RACS architect, S W Ackroyd FRIBA, in co-operation with Frederick Wilkinson, the buildings on the corner of Garratt Lane and Burntwood Lane comprised the chapel of rest and a garage. The chapel was dedicated in October 1933 by the Bishop of Kingston

participation in a 'clothing scheme' to ensure a smart appearance at funerals.

The fleet comprised Daimler and Minerva limousines and hearses, which served the eight branches: Woolwich, Plumstead, New Cross, Well Hall (Eltham), Greenwich, Tooting, Earlsfield and Wimbledon. Others were being planned. The department was built on service:

> 'The desire to be of service is the chief motive which prompts the Executive staff to give every possible assistance to those unfortunate folk in need, and the high standard of efficiency of the RACS Funeral Service is a reflection of Mr Wilkinson's own forceful personality. Mr Wilkinson believes in getting things done, and this characteristic permeates the endeavours of the whole staff when a question of service arises. For the building up of what is recognised throughout the RACS area as an exceptionally dignified, efficient and reverent Funeral Furnishing Service, the Society has to thank Mr FH Wilkinson.' [67]

In October 1935, it was reported that an office had been opened at Rushey Green, Catford, and that two further offices were in the course of construction: Bexleyheath and Walworth.[68]

In January 1936, the RACS started a process of revision of its payment of the death benefit scheme. The society was still paying out a benefit where the member then instructed a private funeral director to manage the arrangements; an estimated two-thirds of members took advantage of this. Rules were passed to have a two-year transition period in which the death benefit scheme would be linked to the funeral-furnishing department, and the benefit would only be payable if the arrangements were carried out by the RACS funeral-furnishing department. [69]

By the 1940s, the Co-operative Society had 10 branches in south London and was carrying out over 2,000 funerals each year. In 1958, this had increased to 15 branches and the 98 members of staff were responsible for nearly 5,000 funerals. It was probably the largest funeral department within the Co-operative movement in the UK.

1. This is confirmed in 'Co-operative Funeral Furnishing. A Review of what Societies are Doing' (1931) *The Producer* December pp362–364. Earlier Co-operative directories cannot be located.

2. See also *Co-operative News* 6 August 1904

3. 'Funeral insurance – An Historical Note' (1998) *FSJ* September p82

4. Carr-Saunders AM, Sargant Florence P and Peers R (1938) *Consumers' Co-operation in Great Britain. An Examination of the British Co-operative Movement* London: George Allen & Unwin Ltd p175,

5. 'Funeral Furnishing' (1932) *The Producer* June p185. See also 'Memorials to the Departed. The New Funeral Furnishing Department of the CWS' (1934) *The Producer* December pp373–374, and 'Co-operative Funeral Furnishing. A Review of what Societies are Doing' (1931) *The Producer* December pp362–364

6. 'Sixteenth Annual Convention of the British Undertakers' Association' (1919) *TUJ* July p194

7. 'Co-operative Societies as Funeral Furnishers' (1926) *TUJ* May p165

8. 'Co-operative Funeral Furnishing. A Review of what Societies are Doing' (1931) *The Producer* December pp362–364

9. 'Distributive Developments: Funeral Furnishing' (1920) *The Producer* August p309. See also September–October (1920) p332. 'Retirement of Mr HD Webster' (1950) *FSJ* October p576

10. 'Co-operative Societies as Funeral Furnishers' (1926) *TUJ* May p165. See also 'Co-operative Societies and the Funeral Trade'(1926) *BUA Monthly* May pp246–247 and 'Co-operative Funeral Furnishing. A Review of what Societies are Doing' (1931) *The Producer* December pp362–364. Author's correspondence with Tony McCarthy and Delma-Rose Yorath. May 2005 and July 2009

11. 'Co-operative Advertising of Funeral Furnishing (1926) *The BUA Monthly* November p108

12. Breck A (1925) 'A Funereal Subject. What About Co-operative Undertaking for Glasgow?' *The Scottish Co-operator* 21 November

13. 'Co-operative Funeral Furnishing' (1931) *The Producer* November p315

14. 'An Expanding Service. Funeral Furnishing 'The Co-op Way' (1947) *The Producer* November pp15–19 and 'The President 1950–51, Mr G Devine' (1950) *FSJ* December p699

15. 'Notes' (1935) *TUJ* September p282. See also 'Funeral Furnishing at Birmingham' (1935) *The Producer* July p222; 'Notes' (1935) *TUJ* September p282; and 'Mr R Whiston joins Ingall's' (1951) *TUJ* May p297

16. 'Funeral Furnishing Trade. Extent of Societies' Interest' (1933) *Co-operative Review* p30–37

17. Carr-Saunders AM, Sargant Florence P and Peers R (1938) *Consumers' Co-operation in Great Britain. An Examination of the British Co-operative Movement* London: George Allen & Unwin Ltd p413. The number of societies with a funeral department differs according to the source consulted. A *TUJ* correspondent writing in May 1934 said that there were 60 towns '... with full funeral facilities, including vehicles, mourning outfits available, as well as cafes at which funeral parties can dine'. 'The Battle with the Co-ops' (1934) *TUJ* May pp147–148

18. 'Editorial: The Co-op's and Co-operation' (1935) *TUJ* April p123

19. 'Co-operative Service in Funeral Furnishing. Rapid Extensions on Federal Lines (1939) *The Producer* April p111

20. Lucas JM (1944) 'Some Practical Questions on Funeral Furnishing *The Producer* June pp15–16, and 'An Expanding Service. Funeral Furnishing "The Co-op Way" ' (1947) *The Producer* November pp15–19

21. 'Co-op Party at Croydon' (1967) *FSJ* July p349. See also Roffey R (1999) *The Co-operative Way* CWS: SE Region

22. 'Outstanding Success in an Essential Service' (1956) *The Producer* June pp17–19

23. 'Funeral Customs in London County' (1927) *BUA Monthly* July pp5–6

24. 'Co-op Undertakers' (1922) *BUA Monthly* August p86

25. 'Sixteenth Annual Convention of the British Undertakers' Association' (1919) *TUJ* July p194

26. 'The Co-operative Societies and the Funeral Trade' (1926) *TUJ* May pp155–156

27. 'London Funeral Workers and the Co-operative Society (1928) *TUJ* September p302

28. 'The Co-operative Societies and the Funeral Trade' (1926) *TUJ* October p338

29. 'The Co-operative Societies' (1926) *TUJ* November pp369–370. See also 'The Co-operative Societies' (1926) *TUJ* December pp401–402

30. 'Co-operative Funeral Furnishing. A Review of what Societies are Doing' (1931) *The Producer* December pp362–364

31. A tax matter was also explored in 1945. See 'Co-operative Societies and Taxation' (1945) *TNFD* November p183

32. ' "Co-ops." An Increasing Danger' (1929) *TUJ* July pp223–225

33. 'The Co-operative Societies Menace' (1930) *TUJ* September p303

34. 'The "Co-op" Menace Demands Right Action' (1931) *TUJ* January p25

35. Ibid

36. 'Editorial Notes: The Anti-Co-operative Movement' (1931) *The BUA Monthly* February pp185–186

37. 'Letters to the Editor' (1931) *TUJ* February pp51–52

38. 'The Co-operative Societies' (1931) *TUJ* March p88

39. Ibid

40. 'Editorial. The Co-operative Societies and the Funeral Trade' (1933) *TUJ* January pp21–22

41. 'The BUA and the Co-operative Societies' (1933) *TUJ* February p47

42. 'Big Developments in Funeral Furnishing by Co-operative Societies' (1933) *The BUA Monthly* December p104

43. 'The Co-ops' (1934) *TUJ* January p13–14. See also 'War with the Co-operative Societies' (1934) *TUJ* January p13 and 'The Co-operative Menace' (1934) *TUJ* February p47

44. 'The Co-operative Menace' (1934) *TUJ* March p82

45. 'The Battle with the Co-ops' (1934) *TUJ* May pp147–148

46. 'Editorial: The Co-ops and Co-operation' (1935) *TUJ* April pp123–124. See also 'The Co-ops and Co-operation' (1935) *TUJ* May pp153–154

47. 'The Disposal of the Dead. Decorum, Dignity, and Efficiency. Aims of the Co-operative Funeral Directors' Association' *The Co-operative News* 13 April 1935 p2

48. 'Co-operative Societies not to be Admitted as Members' (1945) *FSJ* October pp285–286

49. 'Our Readers Write: Vigilance' (1945) *FSJ* September pp253–254

50. 'Our Readers Write: Co-op Monster' (1945) *FSJ* October pp293–294

51. 'Co-operative Societies to be Allowed Membership' (1948) *FSJ* October pp224–226

52. 'Days when Co-operative Funerals were Boycotted' *The Co-operative News* 13 October 1945

53. 'National Co-operative Funeral Directors' Association: A Comprehensive Educational Scheme' (1945) *FSJ* November p319

54. 'National Co-operative Funeral Directors' Assn: Two-Day Conference at Blackpool' (1946) *FSJ* October pp369–371

55. 'Training Scheme for Assistants. Co-operative Funeral Service Managers' Executive meet at Middlesbrough' (1948) *FSJ* November pp341–343

56. 'Training the Young Men as Managers. Co-operative Movement's Educational Scheme' (1948) *FSJ* December pp608–610. See also 'Co-operative Funeral Managers' Conference at Weston-Super-Mare' (1948) *FSJ* November pp552–559 and 'NACFFM: Training for Managerial Appointment' (1948) *FSJ* September pp445–446

57. 'National Association of Co-operative Funeral Service Managers. National Executive Receive Civic Welcome at Rugby' (1949) *FSJ* February p97

58. 'National Association of Co-operative Funeral Service Managers. Reduced Subscription Promised. CWS 100% support for Embalming' (1949) *FSJ* July p319 and 'National Association of Co-operative Funeral Service Managers. Education Section. Week-end School at Newcastle-upon-Tyne' (1949) *FSJ* August p454

59. 'Alexander McLeod's Centenary: Tribute to a Great Pioneer' (1932) *Comradeship and the Wheatsheaf* November p *vi*. Born on 29 September 1832 and died in 17 May 1902, McLeod was buried at Woolwich Cemetery. The statue of McLeod on the Society's building at Powis Street was unveiled on 21 October 1903

60. 'Notes' (1928) *TUJ* September p289. See also *The Times* 10 September 1928 p19. See also Notes' (1928) *TUJ* September p289. See also 'Funeral Furnishing' (1929) *Comradeship and the Wheatsheaf* February pp *iii–iv*

61. 'A Few Words on Funeral Furnishing' *Comradeship and the Wheatsheaf* (1929) January p *xxvi*. See also 'RA Co-op Society' (1929) *TUJ* January p30

62. Wilkinson boxed for England in the Olympic Games in 1918 and 1919 and also wrote about the sport. Much of this biographical information has been derived from 'Trade Personalities' (1936) *TUFDJ* March p91; 'In Praise of Frederick Henry Wilkinson MBIE' (1958) *FSJ* August p373; and 'Obituary. Mr FH Wilkinson' (1965) *FSJ* July p384. See also 'Potted Careers No 5. Mr FH Wilkinson' (1940) *Together* Vol 3 No 2 November pp50–52

63. 'Better Service for Higher Status' (1933) *TUJ* October pp333–336; 'The Psychology of Salesmanship or The Sale of a Funeral' (1934) *TUJ* October pp333 340; 'Funeral Ethics' (1934) *TUJ* March pp77–81; 'Salesmanship of Embalming' (1939) *TUFDJ* August pp281–283. See also 'Funeral Ethics. An Outline of Basic Principles' (1950) *FSJ* April pp191–196

64. 'Funeral Furnishing Department' (1929) *Comradeship and the Wheatsheaf* April p *iii–iv*. See also 'Funeral Furnishing. Societies Should Consider Co-operative Undertaking' (1930) *The Producer* July p211; 'Funeral Furnishing Department' (1929) *Comradeship and the Wheatsheaf* May p *ii*; and 'The Funeral Furnishing Department' (1929) *Comradeship and the Wheatsheaf* November p *iii*. See also 'Funeral Furnishing' (1932) June *The Producer* p182

65. 'The Funeral Furnishing Department' (1929) *Comradeship and the Wheatsheaf* June p *ii*. See also March (1930) p *ii*

66. 'Funeral Furnishing Service' (1932) *Comradeship and the Wheatsheaf* June p *v*

67. 'The Royal Arsenal Co-operative Society's Funeral Furnishing Service' (1935) *TUJ* August p271

68. 'Revised Death and Funeral Furnishing Benefit Scheme' (1935) *Comradeship and the Wheatsheaf* October p *xvi*

69. 'Revised Death and Funeral Furnishing Benefit Scheme' (1935) *Comradeship and the Wheatsheaf* July p *xvii*. See also October (1935) p *xvi*

CHAPTER 8
THE UNKNOWN STORY OF THE UNKNOWN WARRIOR

'The bringing of the Unknown Warrior from his grave in Flanders to the heart of London, where his coming was awaited by King and his Princes, Generals and Admirals, Archbishops and Bishops, Ministers of State, and a multitude of men and women, was the tribute of the nation's soul, and of all great and humbler people made equal in reverence, to the virtue of the Common Man who won the war.'[1]

An important but largely unrecognised event for the British funeral industry was the involvement of the BUA in the bringing home of the Unknown Warrior.[2] Following a suggestion from an army chaplain, Revd David Railton, to the Dean of Westminster, an unidentified body was transported from Flanders for burial in Westminster Abbey. Taking place in November 1920, this event captured the minds of a war-scarred nation. The two representatives of the BUA for this task were the secretary of the London Centre of the BUA, John Sowerbutts, and the BUA national president, Kirtley Nodes. After biographical sketches of the two personalities involved, this chapter comprises Sowerbutts's account of the event, along with recollections provided by Kirtley Nodes and material from other sources.[3]

▼ The coffin was brought into Westminster Abbey to be photographed. The next day it was taken by John Nodes's hearse to Charing Cross station

JOHN SOWERBUTTS

John William Hadrian Sowerbutts was born in East London in August 1875, the second son of Thomas Sowerbutts, a well-known fishing-tackle manufacturer. After being articled to an architect, John Sowerbutts received administrative training in the office of the Clerk to the Guardians of the Whitechapel Union and subsequently became an accountant in the Poor Law branch of the Local Government Board. In 1901 he trained for Holy Orders at St Augustine's College, Canterbury; four years later he was ordained in Calcutta, where he served until returning to England in 1907. He then held curacies in Devon and Essex, and became an Associate of King's College London, but in 1917 he was given permission to take time out from full-time ministry. John Sowerbutts became the assistant secretary for the London Association for the Blind where he proved himself as an able administrator. He was also active in both the Boy Scout Movement and freemasonry. On 11 February 1920 he was elected secretary of the London Centre of the British Undertakers' Association, a full-time position to which he devoted his energy and skills. He held this post until 1926 when he returned to the ministry. His last living was in 1948, when he became vicar of Skillington near Grantham in Lincolnshire. He died on 5 May 1955.

HORACE KIRTLEY NODES

Born in London in November 1876, Horace Kirtley Nodes (always known as Kirtley Nodes) entered the family business of John Nodes at 181 Ladbroke Grove on the death of his father in 1895. He was greatly interested in embalming, having received instruction when Professor Renouard visited London in 1900. He was a member of the BES from its establishment in the same year, a founder member of the BIE in 1927 and its first hon. secretary three years later, a position he held until 1948. In 1930, he toured America with two fellow members of the BIE and was deeply impressed by the facilities, premises and educational opportunities available there. Nine years later, he visited South Africa. He was the BIE's first Fellow and then first Master Fellow. Kirtley Nodes was known to be a great mediator between organisations, and his obituary in the FSJ states that, 'He gave himself unsparingly, to the funeral industry, and embalming in particular, and it may be said that he became one of the industry's most vital personalities, greatly endeared by many.'[4] He was twice president of the London Centre of the British Undertakers' Association, national president of the BUA 1920–1921 and a national honorary member of the

▲ Horace Kirtley Nodes (1876–1959), photographed in 1919. He was known as Kirtley Nodes

◄ Burial of
the dead on
the battlefield

59. THE BURIAL OF TWO BRITISH SOLDIERS ON THE BATTLEFIELD.

NAFD. Away from funerals, he was a member of the Royal Choral
Society and a keen gardener. He died on 2 September 1959, and
was cremated at West London Crematorium.

BRINGING THE UNKNOWN WARRIOR HOME

Wednesday, October 27th
"'Ting, ting, ting!" "Hello!" "Is that Victoria 946?"
"Yes!" "Is Mr Sowerbutts there?" "Speaking!" "Good
afternoon, sir; we are HM Office of Works, and we
have to arrange the burial of the 'Unknown Warrior'.
Can your Association undertake the making of the
casket for the remains? It has to be made to special
designs yet to be got out, and it has all to be done in a
very short time. Can you do it?" "Yes, sir, of course we
can, nothing is impossible to my Association!" "Very
well! Will you arrange a conference with me here at the
Office of Works tomorrow morning at 10.30?" "Yes, sir,
10.30 tomorrow!"

So began the great adventure. The next morning, at King Charles
Street, the design got out by the architect of the Office of Works
was produced. Immediately it was seen that the task, or rather the
labour of honour, was going to tax the Association's abilities to
the utmost. The casket could not be got from any existing stocks
– the furniture was absolutely unique – but still, your humble
servant was confident that the Association could do the work. The
conference was just ending when the National President [Kirtley

Nodes] voiced what all were feeling, viz; that the labour of honour would be marred unless the Association went a step further; accordingly he asked Sir Lionel Earle [Permanent Secretary of HM Office of Works 1912–1933] to accept the whole as a gift of appreciation to the nation for the self-sacrifice of the great body of England's warrior.'

In his account of the discussion, Kirtley Nodes is more precise about the meeting on 28 October:

> '... I well remember that some of the deputation had come armed with most elaborate designs, but we were soon informed they were not wanted. What was wanted was the cost of supplying the casket in accordance with the design got out by the architect of the Ministry of Works, acting under instructions of a special committee of the Cabinet. This was handed to us and explained and we were then asked to go over to a corner of the room and talk the matter over.
> At the time, I was not only a member of the London Centre, but also National President, and I felt that here was something in which the whole of the Association should take part. I proposed to my fellow-members that this should not be a question of competition, but a gift to the nation from our Association as our tribute to the Unknown Warrior. To this they readily agreed and also, to avoid any feeling, the casket should not be made by a member.
> Upon informing the representatives of our decision, he said he felt sure we should say this, although his superior had thought otherwise. As an appreciation of our offer it was decided that two of us should accompany the casket to France, do all that was necessary, and return with it to London. It was felt that by placing the matter in civilian hands, no jealousy should arise between the various sections of the Forces. At the same time we were asked to be diplomatic. The choice of representatives fell upon myself and the Secretary.'[5]

Payment of the casket as a 'gift of appreciation' was funded by inviting BUA members to subscribe one shilling per head and an advertisement was placed in *TUJ*. Nodes commented, 'In this way, I felt everyone could feel they had taken some share. This

To Members.

THE great honour paid to the Association in being asked by the Government to undertake the making of the Casket for the "Unknown Warrior," and our presentation of the same, must still be fresh in the minds of all our Members.

The National Executive is of opinion that it is the desire of all to have some personal share in the gift, and in order that all may have the opportunity, it has been decided to limit subscriptions to 1s. each. Will all Members desiring to contribute either send direct to the Local Secretaries or to the National Secretary.

H. K. NODES,

President.

▸ The appeal to all BUA members for a contribution of one shilling towards the cost of the coffin (*TUJ*)

▾ Ingall, Parsons, Clive & Co coffin factory at Wealdstone where the casket was constructed (*TUJ*)

VIEWS

AT

COFFIN and

CASKET

FACTORY,

WEALDSTONE,
Middlesex.

FACTORY AT WEALDSTONE, MIDDX.

PRIVATE SIDING.

▲ Preparation of the grave in Westminster Abbey
(By kind permission of the Dean and Chapter)

◄ The Unknown Warrior's coffin appeared on the
cover of Ingall's Magazine

suggestion was unanimously agreed to and the money needed
more than subscribed.' Initially, the firm making the casket,
Ingall, Parsons, Clive & Co, who were not members of the BUA,
had offered to donate the casket, but this was declined by Nodes.
News of the BUA's gift soon reached the public domain. The *Daily
Telegraph* noted on 6 November:

> 'The Casket was of special design and had been
> constructed by the British Undertakers' Association, who
> undertook the task free of charge as a tribute to the dead.'

Returning to John Sowerbutt's account:

> 'The Cabinet Committee was sitting that afternoon to
> finally approve the design, and the Secretary was asked
> to remain in his office at Ebury Street until the design
> arrived, so that the work might begin at once, and no
> time lost. It was 7.30pm when the Architect brought
> the design to the office, and that same night work
> was put in hand. It would take too long to tell of the
> succeeding days, the many changes of plan and detail;
> and even when the unique furniture was finished in

just twenty-four hours – the Office of Works was not satisfied with the finish. The design was right, but to get the correct sixteenth-century style it was necessary to send to Carnarvon [sic].'[7]

THE COFFIN

Both John Sowerbutts and Kirtley Nodes give details of the casket's specification. The latter states:

'The casket was to the design of a 16th century treasure casket, made of two-inch English oak, with slightly rounded sides and lid, and finished with a dark wax finish. It was mounted with four pairs of hammered wrought-iron handles and iron bands from head to foot and across the shoulders. The plate was to the pattern of a 16th century shield and bore the following inscription in Old English:

a British Warrior
who died in
the Great War
1918
for King and Country[8]

This was bolted to the lid over a 16th century Crusaders' Sword, given by HM King. The ironwork was made in London, but, in deference to the wishes of Lloyd George, sent to Carnarvon [sic] and finished off by the man who made the ironwork of the investiture of the Prince of Wales.'

The source of oak was the grounds of Hampton Court Palace, and the inner shell was made from one inch of English pine. The casket was made at Ingall, Parsons, Clive & Co's Forward Works in Mason's Avenue, Wealdstone in north-west London. Walter Jackson was responsible for its construction. The ironwork was made by D J Williams at the Brunswick Ironworks, Caenarfon.

Returning to Sowerbutt's recollections:

'A little more than seven days and the task was complete, admittedly a beautiful piece of work and admired by all who saw it. At the last minute, the Office of Works asked for a [coffin] shell to be sent up to the battle front. This was provided in a few hours and despatched to France.

Sunday, November 7th, saw the casket deposited in the Abbey to be photographed amidst architectural surroundings. It reposed in the Jerusalem Chamber until Monday morning the 8th.'[9]

The casket was conveyed to Westminster Abbey in John Nodes's Rolls Royce hearse. A photograph in the Abbey library shows the motor hearse parked outside the West Front and the casket being loaded onto the bier. A nameboard with 'John Nodes' is clearly displayed in the side compartment. The casket would then have been driven the short distance to Charing Cross station for loading onto the train to France.

Sowerbutts continues:

'It [the casket] was then taken to Charing Cross and placed in a special coach and entrusted to representatives of the BUA, and taken to Boulogne. The London Secretary and the National President were authorised to bring the remains to England, and to deliver them to the Military at Victoria, on Wednesday evening. After the casket was locked in the coach at Charing Cross, the keys were handed to the Secretary, with the following official authorisation:-

HM Office of Works
4th November, 1920

TO ALL WHOM IT MAY CONCERN:-
The bearer of this letter, Mr John W H Sowerbutts and his colleague, Mr H K Nodes, have been authorised to accompany the casket to contain the remains of the "Unknown Warrior" to France, and having sealed it, to accompany it back from France to this country, and hand it over to the Military upon its arrival at Victoria on Wednesday, the 10th instant.
(Signed) LIONEL EARLE
Secretary

Eleven o'clock saw us off. Reaching Folkstone Junction at 1.45, a special coach was attached to the casket coach, and the two were run down to the Harbour. It was 6.30pm before we got the Treasure Chest aboard the "Invicta."

At 7pm, we were away – it was a delightful passage across the sea – all that could be desired. Fog was run into off the French coast, but all was well, and soon we were alongside the jetty at Boulogne. The "open sesame" of OHMS saw us through the customs and other formalities, and, the casket safe under Military Guard for the night, we fetched harbour in the Hotel de Londres on the harbour front.

Tuesday, 9th – The car has just come, and with the casket following in an ambulance, we are off to the hills at the back of the town and to the grim chateaux. Just picture the quadrangle of an old French fortress. We pass into it through two low archways; on the left, through a low doorway, we enter a small room draped with bunting and dressed with greenery. The floor is strewn with laurels and chrysanthemum blossoms; this is the Chapel of Repose. Outside is drawn up a guard of honour poilus; here and there are groups of other officers, French and English, and so we await the coming of "our boy".

Presently one or two staff cars swing into the quadrangle and then just a battle-scarred ambulance, the familiar red cross on its sides, and before we can realise it, is borne on eight shoulders the simple [coffin] shell covered with England's flag [The Union flag], and within it the "Unknown Warrior". It is a touching scene. Gently we lay him in the "Treasure Chest". The lid is on, the bolts are driven home, and we step forward wondering "whose is this image and superscription" – and there, on a shield of hammered iron, we read: "A

▼ The coffin being carried from the chapel ardente in the old chateau of the port of Boulogne

THE UNKNOWN WARRIOR LEAVING FRENCH SOIL FOR HOME.

British Warrior who died in the Great War, 1914–1918, for King and Country." Bound round the waist and from head to foot in bands of hammered iron, with a crusader's sword thrust under the shield, we leave him in the guard of France for the night, but in the casket of England's heart, bound round with the bands of grateful love. Tomorrow we will bring him home on a destroyer, when the French have done him their devoir. Midnight, Tuesday – The hotel has become temporary Staff Headquarters. There is a field telephone on the table, the lounge is full of staff officers, English and French; soldiers and officers are coming and going, all busy preparing for tomorrow's function. All the chiefs are staying there to-night. The French mean to make a big show. We have just learned that the Destroyer is timed to arrive at 10, and to leave at noon.[10]

Wednesday the 10th, 10am. A dull morning, the harbour is thronged with people, the roadways and quayside are beflagged; presently, out of the mist over the sea, a grey form emerges, it is HMS Verdun, upon which "our boy" is to go home. It does not take long to make fast alongside. After certain preliminaries the Commander comes ashore and stepping towards the representatives of the BUA, invites them aboard. Then a period of waiting, until in the distance the sound of French trumpets, playing the stirring "Aux Champs", and the deeper, impressive music of the massed bands

▼ The Unknown Warrior on HMS *Verdun* returning to England

with Chopin's "Marche Funèbre". Slowly down from
the chateau, where he had rested all night under the
vigil of a myriad November stars, to the Place de
Dernier Sou, from thence with General Sir George
Macdonagh and Marshal Foch following, borne in
a French Army Service wagon, through bareheaded
ranks to the quayside.

It is a scene never to be forgotten, absolutely unknown,
just a skeleton and no more. It might have been a baby
in the shell for the weight; yet he is a British Warrior.
With the grandeur and triumph of a great victor he
comes aboard and, resting on the quarter deck, abaft
two grim torpedo tubes bowered in flowers and veiled
in battle-stained gown of three crosses intermingled,
we make him fast for the last crossing. It would take
a volume to describe it all. Shortly we cast off and out
into the "glory mist" we glide. Just outside the harbour
six British destroyers file into position, three each side,
and four French cruisers meet us with their salutes,
whilst the guns of the land forts boom au revoir, and
the bells of Notre Dame sing a requiem.

About six miles out the French cross our bows firing
their farewell, and so we come onwards. About 2pm
we sight "Blighty", and a wireless message is received
asking us to cruise in the Channel, arriving at Dover
at 3.30; accordingly we go west with our escort. At 3.25
we enter Dover harbour, our escort falling away to right
and left, the fortress calling to us with nineteen guns.

Never can one forget the scene. We just glide in and,
as we get close to the jetty, the bands sing softly, "Land
of Hope and Glory." Soon he is borne aloft on his
comrades' shoulders, and we follow him ashore to the
garden of repose; for although it is a prosaic railway
van, yet it has made history before; for in the same van
they brought home Captain Fryatt and Nurse Cavell.[11]
Beautifully draped, garlanded and banked with laurel
and chrysanthemum, it is just a paradise. We lay him
down, and until the train starts the doors are left open
for the public to view. At 5.30 we are away and all the
way up at every station are crowds; by the crossings,
on the bridges out of the darkness, white faces peer
as we speed with our treasure to London. The rest the
papers will tell you. Suffice it to say that at Victoria

the BUA representatives handed the keys of the coach and its precious treasure to the Officer Commanding the Guard, and received the following receipt for the same: -

Received from Mr Sowerbutts, two keys of Van no 132, SE & CR

(Signed) PS GREGSON ELLIS
Lieutenant
1ˢᵗ Battalion Grenadier Guards
November 10ᵗʰ, 1920

So ended a work of which the BUA must ever be proud, and one which reflects great honour upon the Association; and so far as the two representatives were concerned, a labour of love they will never forget. Requiescat in pace – Gloria in æternum.'[12]

▼ The unveiling of the Cenotaph as the casket of the Unknown Warrior passes

▼ The grave in Westminster Abbey

AFTERWARDS

Sir Lionel Earle was prompt in writing to thank the undertakers. His letter written from HM Office of Works in Westminster on 12 November 1920 is still in the Nodes family's possession. It reads as follows:

> *Dear Sir*
>
> *I desire to take the earliest opportunity to convey to you the warm thanks of His Majesty's Government for the assistance which you rendered in taking the Casket to contain the remains of the Unknown Warrior to France, and in accompanying it back to this country.*
>
> *On all sides I have heard nothing but praise for the admirable way in which the duties entrusted to you were carried out.*
>
> *HK Nodes Esq*
>
> > *I am, Sir*
> > *Your Obedient Servant*
> > *Lionel Earle*
> > *(Signed)*

Although John Sowerbutts and Kirtley Nodes gave accounts of their role in the events, both were anxious to emphasise its importance for the funeral profession. Writing in December 1920, the *TUJ* editor interviewed Kirtley Nodes, where he commented:

> '*I don't want ... to make too much of the bringing home of the 'Unknown Warrior'; but the honour done to the BUA in that matter is an asset of no small value ... We did not take the credit ourselves. It was the BUA they were honouring.*'[13]

At the BUA conference in Scarborough the following year, the subject was raised (it was Kirtley Nodes' presidential conference; John Sowerbutts was also present). *The BUA Monthly* reported:

> '*The Association was called upon by the Government to take part in the burial of the "Unknown Warrior". It was their privilege and pride to do so. They counted it as a great honour to take part in that ceremony and to present the casket used, and it was [Nodes and Sowerbutts] privilege to go to France and collect the*

remains. Mr Nodes alluded to the solemn and imposing character of the ceremony to bringing the body to England and its subsequent burial at Westminster Abbey. The touching appeal of his speech made a deep impression on the audience, and many wet eyes were in evidence.'[14]

▸ A replica of the coffin of the Unknown Warrior at Bodium on the Kent and East Sussex Railway. It is displayed in the restored van that conveyed the coffins of Edith Cavell, Captain Fryatt and the Unknown Warrior

▾ The BUA national president, Kirtley Nodes, with members of the executive laying a wreath at the Cenotaph in May 1921 (*BUA Monthly*)

1. Gibbs P (1920) 'A Soldier Known Unto God' *The Illustrated London News* 20 November

2. See Gavaghan M (2006) *The Story of the British Unknown Warrior* Fourth Edition Le Touquet: M&L Publications;, Wilkinson J (2006) *The Unknown Warrior and the Field of Remembrance* London: JW Publications; and Janes B (2012) *The Unknown Warrior and the Cavell Van* Tenterden: Kent and East Sussex Railway. See also 'How the Unknown Warrior was Selected' (1939) *TNFD* December p219

3. The main text is taken from Sowerbutts J W H (1920) 'Bringing the "Unknown Warrior" Home' *TUJ* November pp340- 341

4. 'Obituary: Mr H Kirtley Nodes' (1959) *FSJ* October p452. See also 'The Late Mr H K Nodes: An Appreciation' (1959) *TFD* October p208

5. Nodes HK (1944) 'Bringing the "Unknown Warrior" Home: Mr H K Nodes Recalls NAFD's Part in Historic Event' *TUFDJ* June pp173–174

6. The *Daily Telegraph* 6 November 1920

7. Sowerbutts (1920) p340

8. Nodes (1944) p173

9. Sowerbutts (1920) p340

10. Sowerbutts (1920) p340

11. For Nurse Cavell see page 300. Captain Fryatt was executed by the Germans in 1916 for trying to ram a U-boat. After exhumation in 1919, his coffin was brought by boat to Dover and then by train to Victoria.

12. Sowerbutts (1920) p341

13. 'Interview with the National President, BUA' (1920) *TUJ* December p393

14. 'The 1921 Annual Conference at Scarborough' (1921) *BUA Monthly* July p19.t

CHAPTER 9
THE KITCHENER CASE: THE EXTRAORDINARY STORY OF AN EMPTY COFFIN

Whilst the burial of the Unknown Warrior, as revealed in the previous chapter, was an event involving the funeral industry, an astonishing hoax that duped so many was in 1926. This chapter examines the extraordinary story of Lord Kitchener's remains.

On the night of 5 June 1916, HMS *Hampshire* set sail from Scrabster, north of Thurso in Scotland, on its way to Russia. The secretary of state for war, Lord Kitchener, was on board along with his staff, government officials and a 650-strong crew. The purpose of the trip was to discuss the purchase of munitions to be used in the war effort against Germany. At around 8 p.m. the ship struck one – possibly two – mines off the mainland between Brough of Birsay and Marwick Head. An electrical fault prevented the launch of lifeboats and the *Hampshire* sank within 15 minutes. Only 12 crew members survived. Lord Kitchener's body was never recovered.

The tragedy gave rise to much speculation that Lord Kitchener, disliked by the War Office and politicians, had been sent to his death. Conspiracy theories soon emerged, for example that this act of sabotage had claimed a person impersonating Lord Kitchener and that there were sightings of the real Kitchener in London's Whitehall, and also in Washington, Cairo, Rome and Cyprus. An official enquiry was demanded into the sinking of the *Hampshire*, but this never took place and the myths continued.

▼ Earl Kitchener of Khartoum (1850–1916)

▼ The explosion aboard HMS *Hampshire*

Thy Will be Done.

SEE round Thine ark the hungry billows curling;
See how Thy foes their banners are unfurling;
Lord, while their darts envenom'd they are hurling,
Thou canst preserve us.

In 1921, a film was made entitled *How Kitchener was Betrayed* (it was not well received by members of Parliament, officials from the War Office or Kitchener's sisters). Among the audience at the private screening was a journalist called Frank Power (whose real name was Arthur Vectis Freeman). He was taken with the idea made by the film's agents that Kitchener's body could be found in a grave in Norway. In 1926, Power took the step of accusing the Admiralty of Kitchener's assassination, in articles and two books: *Is Kitchener Dead?* and *The Kitchener Mystery*. In these, he maintained that a monstrous betrayal had taken place. Furthermore, Power claimed that it was well known that Kitchener was on the *Hampshire* and therefore an easy target for the Germans, and that there had been two explosions on the ship with the second being from *inside* the decks.

Power claimed it was possible to return Kitchener's body to England for burial. Writing in *Is Kitchener Dead?* he stated:

> *'I am going to Norway to prove in the most conclusive possible way that Lord Kitchener is dead, but I shall still hope for the final vindication of the great soldier,*

▼ The cover of one of Frank Power's books outlining his theory

▶ A map showing the place of explosion on the ill-fated journey published in *The Kitchener Mystery* by Frank Power

THE KITCHENER
MYSTERY

By FRANK POWER

Price Two Shillings

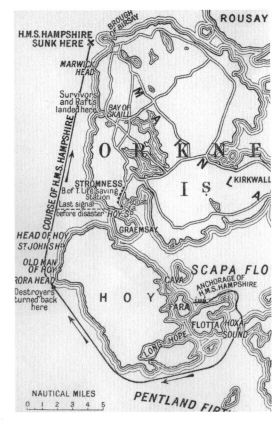

to which his sacrificial passing alone entitled him a
last resting place in Westminster Abbey; and, without
rancour and without hate but simply as an act of
British justice, I shall hope for the final confusion of
those whose hands have struck in the dark.'[1]

Supported by the British Legion and publicity from the screening of another film, *The Tragedy of the Hampshire*, Power left England for Norway. On 12 August 1926, *The Orcadian*, the Hebredian weekly newspaper, announced that Power had located the grave. He wrote:

'In my quest I received invaluable assistance from
two of the actual men who found that body on the
coast. It was they who first of all gave it temporary
burial beneath the stones on the shore itself, whence
it was afterwards transferred, in a wooden casket, to
the cemetery. These men have both of them given me
signed statements which I have set down in full, so
that there cannot be any shadow of doubt as to their
authenticity.'[2]

Power then detailed the exhumation:

'After I had taken precautions against any untoward
disturbance or irreverent curiosity, my four Norwegian
helpers removed the few feet of soil above the wooden
shell in which Lord Kitchener's body has been put
after its temporary interment by the fisherman on
the seashore. This covering was very much decayed,
though holding together.

With care and reverence, and after reading some passages from a beautifully illuminated form of prayer specially arranged for my "Pilgrimage" by a lady who was an intimate friend of the late Earl, I moved the pieces of wood which covered all that remains of Lord Kitchener.'[3]

Power organised for a photographer to record the events whilst a cameraman working on behalf of Crystal Productions Limited of London filmed the proceedings. A mock funeral procession was held at Stavanger, Norway, where the remains were carried on board a ship destined to Southampton. From there the coffin was taken on a mail train to London.

Mr Rudd   *Nov 17 1920*

DR. TO T. HURRY,

Complete Funeral Furnisher,

164 WATERLOO ROAD, LAMBETH, S.E. 1.

——— AND ———

868 GARRATT LANE, TOOTING, SURREY.

Telephone No. : 1877 HOP.   ESTABLISHED 1868.

Power arrived back in London with what was described as a 'long packing case'. Using the name of 'Fraser' – the name of the consignee – he telephoned a funeral director, T Hurry, with instructions to remove the coffin (weighing 4cwt 7lbs), to his premises. Thomas Hurry's premises were near the station at 164 Waterloo Road. Staff arrived in a motor hearse only to be charged a £4 levy as the packing case had been identified as a coffin by a vigilant porter. The coffin was subsequently taken to his chapel of rest and Edward Ackril, the owner of T Hurry, awaited further instructions. Up until this stage, all Power had told him was that the case contained the body of a man.

In the meantime, Power had written to Stanley Baldwin, the prime minister, to inform him of his intention to bury the body of Lord Kitchener in Westminster Abbey:

> *'Sir, Following my previous communication to you, to which, however, I had only a formal acknowledgement, I beg to inform you that I have in my possession the remains of a person which I am convinced can be identified as those of the late Earl Kitchener of Khartoum. As the head of the Government I hasten to tell you that they are at the disposal of the competent department in this matter, and I should be only too ready to keep any appointment to carry out any suggestions that you may make.'[4]*

▲ A letterhead dated 1920 showing the premises of T Hurry on Waterloo Road

◄ (L to R) Dr Samuel Ingleby Oddie, the coroner who examined the empty coffin. Sir Bernard Spilsbury who was also present when the coffin was opened

▼ The empty coffin (*BUA Monthly*)

However, there was a stumbling block – no certificate existed to permit burial. The only way to obtain one was to refer the matter to the coroner. Officers from Scotland Yard then visited Hurry's premises and removed the coffin to the public mortuary in Lambeth High Street, where it rested overnight.

At ten o'clock on 16 August, Dr Samuel Ingleby Oddie, the coroner for the City and Liberty of Westminster and also South Western District of London; along with Chief Constable Frederick Wensley; the Honorary Pathologist to the Home Office, Sir Bernard Spilsbury, with his assistant, Hilda Bainbridge; and several officers from New Scotland Yard attended the mortuary.[5] A large crowd had gathered outside. *The Orcadian* provided a detailed account of the events at Lambeth:

> '*The mortuary keeper took off the lid of the packing case. The nails were easily forced. Inside was a large chest of bright copper. It enclosed the coffin, but was not fastened to it. The copper case removed and (according to a Home Office statement) the coffin was found to be of unpolished elm and new. The lid was unscrewed, and then there was the discovery that the coffin contained nothing. The screws were put back, and the two were taken off to Scotland Yard in the police tender.*'[6]

One source states that the coffin was ' ... unevenly spread with fresh tar.'[7] This would not be unusual as at the time pitch was used to seal the interior of a coffin. Detectives then made for Sinclair Gardens in Kensington where Power lived, to discuss the matter. He was not at home. When, eventually, he was caught up with he said:

> 'I have a clear conscience in the matter. I am not sheltering behind the law and I have not broken any regulations, nor made any false declaration. Neither have I hoaxed anybody ... I am going straight to Scotland Yard ... to demand that I shall be given an opportunity of examining the inner shell of the coffin ... If only I can look at that inner shell I shall have something to say on the subject. I do not dispute that the polished outer coffin or the copper casket are those that I brought from Norway, but I marked the inner shell of elm in such a way that I can at once tell it if has been removed and replaced by an empty one.'[8]

He added:

> 'I had asked Scotland Yard if I could be present when the case was opened, and was told that it would be quite unnecessary, and I need not trouble myself ... I considered it was only right that I should be there when the case was opened, and I went down to the mortuary to see if they would admit me. I went up to the officer at the door and said 'I am Frank Power.' He laughed and replied, 'That be hanged for a tale (sic). There have been a dozen Frank Powers here already. You can't go in.' So I gave it up as a bad job, and left with my wife by the 11.50am train to Portsmouth.'[9]

Power declared that Lord Kitchener's remains had been removed from the coffin at some point between being taken from Hurry's premises to the mortuary.

In quoting from *The Daily Chronicle*, Joanna Bourke sums up the outrage:

> 'The press expressed the widespread fury. The hoax was said to have poured scorn not only on the sentiments of Kitchener's family, but on thousands of

men and women who cherished a "deep veneration of the memory of the great soldier who had sacrificed his life to the country's need during the war".[10]

The Observer described it as a 'vulgar burlesque' whilst a leader in *The Manchester Guardian* said:

> *'The legend that Lord Kitchener still survives, or that he died a prisoner in Germany long after the sinking of the Hampshire, seems almost a rational exercise of the human faculties compared with the tale of the coffin which, as the Home Office declared yesterday, was found to contain no body at all. The one set of fables is a consistent and unsupported myth; the other is a mysterious farrago into which enter such confusing realities as a real but empty coffin apparently brought all the way from Norway, a coroner, a Government pathologist, and preparation for an inquest on a body which did not exist. One looks in vain for the least evidence of a rational motive which would pull this jumble into even an understandable delusion. The thing is now described as "a hoax", but even hoaxers prefer to avoid an essentially disgusting hoax which is certain to be found out. The whole thing sounds like an essay in sensational fiction, but in sensational fiction so badly constructed that it is all details and no attempt at a plot. The intelligent newspaper reader can only shake his head over it and pass on. But what will happen with the unintelligent newspaper reader can*

▶ The effigy of Lord Kitchener in St Paul's Cathedral

be but dimly imagined. As the story stands at present, it looks as though in this case it might very well become the centre of some even more preposterous "Kitchener myth" than has yet been spread from one wild head to another.'[11]

Despite its extraordinary nature, Dr Oddie does not mention this event in his autobiography, but the reminiscences of Frederick Wensley does contain a reference.[12] The chief constable recognised that Power had not committed a criminal act; nevertheless, he had felt that ' ... there was an indecency in the proceedings and that the thing should be put a stop to.' He wrote:

'On thinking the matter out the solution flashed across my mind. Those responsible for the affair asserted that the coffin contained a body. "Very well," I said. "Where is the doctor's certificate showing the cause of death?" Of course there was no certificate. "Then," said I, "this body must be removed to the mortuary to await a coroner's inquest." This exploded the whole silly story at once, for it enabled the coffin to be opened and examined. There was no body there!'[13]

On 9 September 1926 the Home Office issued a statement about the matter, which was widely reprinted in the newspapers. The investigation recounted the background, then focused on the coffin:

'The exhaustive inquiries which have just been completed established conclusively that this coffin has never left the country. It was purchased by Mr Power in Kirkwall, early in 1926, and remained in store in Kirkwall until July 27 last, when, on Mr Power's telegraphic instruction, it was despatched to him in Newcastle. Mr Power, who had left for Norway on July 20, returned to Newcastle on August 4, and ordered the coffin, which had meanwhile been stored in a Newcastle warehouse, to be sent to him in London. The movements of the coffin at every stage, from the time it left Kirkwall until the time of its opening in London, have been completely accounted for, and the coffin has not been tampered with in any way in transit.'[14]

The statement concluded:

> 'As a result of the inquiries which have been made, the public may therefore be satisfied that the statements that the remains of the late Lord Kitchener in Norway, disinterred, and sent to this country are entirely without foundation.'[15]

However, it was the *BUA Monthly* that provided the most authoritative account of the sequence of events:

> 'Mr Samuel Baikie, Bea Cottage, Stromness, was interviewed by a representative of *The Orcadian*. He stated that the coffin purchased by Mr Frank Power, which has been stored in the yard of Messrs S Baikie & Son, wood merchants, Kirkwall, was put to Kirkwall pier on the afternoon of 26 July. It was enclosed in a large packing case, addressed to Mr Frank Power at Newcastle-on-Tyne. It left Kirkwall the following morning per the North of Scotland and Orkney and Shetland Steam Navigation Company's steamer en route for Newcastle via Leith.
>
> Mr Baikie has no further knowledge of the matter beyond that the account for the freight came to him, and as the bargain with Mr Power was that the latter was to pay freight, that account was sent to the purchaser for settlement. He had never at any time since he despatched the coffin had any inquiry in regard to the matter, nor had he had any communication from Mr Power.
>
> Mr Baikie's attention was drawn to the statement by the Home Office that the coffin was quite new. He replied that it was certainly in new condition, but that to anyone with any knowledge of wood, it could clearly be seen that the coffin which he sent to Newcastle had been made some considerable time before, as the elm case was showing distinct signs of seasoning. Otherwise both the outer elm and the inner shell of copper looked in absolutely new condition.
>
> Asked when he removed the coffin from Lyness Naval Cemetery Chapel, Mr Baikie said he could not put an exact date on it, but, speaking from memory, he gave as an approximate date August 1925.

We enquired if Mr Baikie obtained the coffin along with surplus Naval stores he purchased at Lyness. He said that was not exactly the position. A number of coffins which were lying at Lyness came on the market. All the others were black-covered and were purchased by another party, who, however, did not require the one made of elm and copper lined. Mr Baikie was approached on the matter and purchased it. Up to that time, this coffin had always been known as "Kitchener's" coffin.[16]

As the chief constable noted, Frank Power had not committed a crime and therefore was not guilty of any offence. Like Kitchener, he disappeared without trace.

1. Quoted in Bourke J (1996) *Dismembering the Male: Men's Bodies, Britain and the Great War* London: Reaktion Books pp237–242

2. *The Orcadian* 12 August 1926

3. ibid.

4. *The Orcadian* 19 August 1926

5. Rose A (2007) *Lethal Witness. Sir Bernard Spilsbury, Honorary Pathologist* Kent (Ohio): Kent State University Press

6. *The Orcadian* 19 August 1926. See also *The Times* 16 August 1926

7. Bourke (1996) p241

8. *The Orcadian* 19 August 1926

9. ibid.

10. Bourke (1996) p241

11. *The Manchester Guardian* 17 August 1926

12. Oddie S I (1941) *Inquest* London: Hutchinson

13. Wensley F P (1931) *Detective Days: The Record of Forty-two Years' Service in the Criminal Investigation Department* London: Cassell p214

14. *The Times* 10 September 1926

15. ibid.

16. 'The Kitchener Myth – Its Sequel' (1926) *BUA Monthly* October p70. See also 'The Kitchener Myth. What We Know of it' (1926) *BUA Monthly* September p56.

CHAPTER 10
THE WORLD'S GREATEST AIR TRAGEDY: THE R101

▲ The R101

▼ The crash scene at Bois des Coutumes near Beauvais

THE DISASTER

At 6.24 p.m on Saturday, 4 October 1930, the R101, the largest dirigible ever constructed, left the Royal Airship Works at Cardington airfield in Bedfordshire destined for India. On board were 54 passengers and crew.

The airship crossed safely into France, but shortly after 2 a.m. it struck ground on a hillside at Allone, near Beauvais, and was engulfed in fire. Of the 46 who died in the crash, among them was Air Minister Lord Thomson; Sir Sefton Brancker, the director of civil aviation; and Major G H Scott, assistant director of airship development at the Air Ministry and officer in command of the flight. A further two died of their injuries three days later.[1]

A funeral director in London was immediately instructed to arrange for the recovery of the bodies and to assist with the preparation for the funeral in London, and accounts of the involvement appeared in both the *Undertakers' Journal* and the *BUA Monthly*. Supplemented by additional research, the extracts from these articles form the basis of this unique insight into the events immediately after the crash, concluding with the burial.

The firm of Mills & Co (National Reformed Funerals Company) of Paddington was appointed 'Technical Advisor' to the Air Ministry and was responsible for the recovery of the bodies, assistance with the arrangements in London and preparation of the grave. It was a task completed in six days.[2]

THE RECOVERY OF THE VICTIMS

Immediately following the Air Ministry's request to assist, staff at Mills & Co made contact with a Paris-based funeral director and embalmer, Bernard J Lane.

Known as the 'American' funeral director in the city, Lane was in fact born in England but moved to Canada; it was while he worked as a funeral director that he became interested in embalming. He studied at the Cincinnati College of Embalming, then moved to Vancouver to join the Canadian Expeditionary Forces in 1914, and in France served with the Canadian Machine Gun Corps. He then worked with E Teysseyre, a Parisian funeral-directing business, and later married the daughter of the proprietor. He settled in France but regularly visited England, and often attended the annual British Undertakers' Association conferences.

Bernard Lane provided accounts of his work on the R101 disaster to a representative of the *TUJ* :

> *"On Sunday morning I received a telephone message to proceed to the wrecked airship as the representative of Messrs. Mills & Sons, of Paddington, who were instructed by the Air Ministry to remove the victims to London ...   I left Paris in my car and arrived at 12 noon, finding Major Neville, Military Attaché to the Paris Embassy, who was the first British officer to arrive on the spot, and I was told that Wing-Captain Bone, the Air Attaché, was expected to arrive very shortly."*[3]

An article published in the *BUA Monthly* reveals the hazardous recovery operation:

> *'The aftermath of the catastrophe was similar to the work of a burial party on active service, combining as it did all the horrors of a battlefield, and the sad sight of the human debris that was extricated from the wreck was one to try the stoutest heart. At an early hour on Sunday, the military had thrown a strong cordon of troops about the gruesome scene, while brass-helmeted firemen fought their way through the wreckage with blow-pipe and torch. Mounted orderlies were dashing to and fro bearing orders and messages (as there were no other means of communication with the base of supplies) and squadrons of French aeroplanes flew*

▲ Bernard J Lane in his Paris office

▶ Placing the bodies in temporary coffins for transportation to Allone town hall

overhead in homage to the dead and dying. A regiment of infantry with detachments of colonial cavalry had been mobilised and the gendarmerie and garde republicaine patrolled the roads, directing the heavy traffic which soon congested the highways.

"The placing of the charred bodies in temporary coffins and removing them from the muddy field where the giant aircraft came down was soon under way and went on all day Sunday. The following morning, Mr Lane received from the Ministry of Air orders to proceed with an attempt to identify those bodies that still bore evidence of any kind, after which they were transferred to the permanent coffins."[4]

It is unclear if any embalming was carried out by Bernard Lane; indeed, a number of factors indicate that this is unlikely: many of the bodies would have been severely burnt; they were all being swiftly repatriated to England; the time-frame did not permit treatment by one person; and, being autumn, the weather conditions would not have aided decomposition. Furthermore, there is no mention of embalming in any accounts of the recovery.

The bodies were taken to the town hall in Allone where Bernard Lane arranged for the delivery of coffins suitable for transportation to England. He continues:

> "'The remains of the victims had already been removed from the wreck by the firemen of Allone, the small village nearby. I immediately communicated with the local undertaker at Beauvais, who sent me out rough coffins as soon as his small staff could make them up. As these coffins were delivered we removed the bodies to the Town Hall of Allone, which had been converted into a 'Chapelle Ardente' for the occasion by the local municipality. By 6.30 on Sunday evening, just before darkness came on, the last of the bodies were removed. The wreck was smouldering as we walked away and flames could still be seen.
>
> "Before I left Paris on Sunday morning, I arranged with a large coffin manufacturer to rush up to Beauvais, on Sunday night, the forty-seven oak coffins with lead shells. On Monday morning we proceeded with the work of identification of the bodies. This was very difficult as in most cases the bodies were so badly burnt, and it was only possible to identify them by any clothing or other articles found on them, such as watches, rings, cigarette-cases which were engraved with names or initials. I am pleased to say in this way we were able to positively identify five of the bodies. The remains were then placed in the oak coffins, which were numbered, and the effects put into small boxes with the corresponding numbers. The coffins were moved to the Town Hall of Beauvais, which had likewise been transformed into a Chapelle Ardente. Forty-seven Union Jacks were specially rushed over from London by Messrs. Mills & Son and placed on the coffins.'" [5]

TRANSPORT FROM FRANCE TO ENGLAND

Although Bernard Lane probably had little hand in the arrangements to transfer the coffins to the railway station on Tuesday, details of the scene were included in the *BUA Monthly*:

> "'Most impressive were the solemn funeral rites accorded by the French Government. A special

chapelle ardente was installed in the medieval Hotel de Ville of Beauvais, and there high officials of many of the allied nations came to pay their respects to those who had sacrificed their lives in the ill-fated mammoth of the air when it crashed in flames and exploded. The ceremony of transferring the bodies to the special train for the channel port was carried out with military precision and with all the martial pomp and panoply that accompanies a state funeral. No more imposing or grandiose funeral procession has been seen in France since that of Ambassador Herrick or Marechal Foch, and it is proposed that a commemorative medal be struck as a pious souvenir of the sad occasion. A salute of 101 guns was fired by four brigades of artillery, while the passing-bell of the historic cathedral sounded the death knell, and at Boulogne a final salvo was fired by three French warships, specially detached from the fleet to pay this last tribute on quitting French soil. The mourning of the townspeople where the ship came down was as sincere as it was profound, and not a flower remained in the whole countryside, all having been gathered into sprays and wreaths which were offered in token of the genuine sympathy that went out to those men whose devotion to duty is their lasting glory."

Thousands of people watched in the rain as the two-mile long procession left the Place Jeanne Hachette in Beauvais. Covered by Union Jacks and surrounded by wreaths, the coffins were conveyed on long-based carts with slatted sides, drawn by horses. Troops lined the streets, while 40 aeroplanes circled overhead and dropped flowers. France observed one day of mourning. Bernard Lane continues:

"'On Tuesday afternoon, the coffins were conveyed from the Town Hall to Beauvais Station, military honours being rendered by the French troops. The coffins were then loaded on to a special train, which immediately left for Boulogne, where we arrived the same afternoon. Military honours were again rendered and the coffins placed on the two destroyers which had arrived to convey the remains to England. Just before the time of departure it was discovered that

*one destroyer had damaged a propeller in getting into
the harbour, so the coffins were transferred to the other
one, appropriately named 'Tempest.' We finally got
under way at 7.30 p.m. and arrived at Dover at 9.30,
after an extremely rough passage, which was quite an
experience to me.*

*"Owing to the strong tide there was some delay in
getting moored. However, we finally got the coffins
disembarked after honours had been rendered by the
crew of the ships and the R.A.F. detachments. The
coffins were then loaded on to a special train, the
coaches of which had been appropriately draped in
purple with carpets on the floors..."*[7]

The coffins were transferred by RAF bearers to a special train
comprising seven vans to accommodate the coffins. Departing
from Dover Marine station, the train arrived at Victoria at 1.25
a.m. Although night time, many people stood at wayside stations
to salute the train on its way to London. Three of the crash
survivors were passengers on the train.

From Victoria station, the coffins were transported in a
circuitous route on 24 RAF tenders to the coroner's mortuary
in Horseferry Road, Westminster, where the procession arrived
around 3 a.m. Even at this early hour, thousands of people lined
the streets around Victoria and Pimlico.

THE CORONER

The Times noted that the French had:

> *"" ... waived all formalities on the understanding that
> the usual forms would be observed in this country.
> This leaves the British authorities responsible
> for identification and investigation. To assist in
> identification personal belongings found near all the
> bodies have been carefully preserved and labelled.
> Relatives will probably be able to satisfy themselves
> with the help of these relics, in which case there will no
> necessity to open the coffins. Should that need arise,
> however, a mortuary is the suitable place for the sad
> offices associated with identification."*[8]

With no certificate available to permit burial, the deaths were
referred to the coroner for Westminster and south-west London,

Dr S Ingleby Oddie. A naval surgeon before becoming a barrister, he had always had one ambition: to become a coroner. Oddie succeeded John Troutbeck as coroner in 1912 and retired in 1938, during which time he held around 20,000 inquests. In his autobiography, Oddie mentions his involvement with the victims of the R101:

> *"As the persons had all died violent deaths and were lying in my jurisdiction, it became my duty under the Coroner's Act, 1887, to hold an inquest to issue the necessary burial orders, and to send my certificates of death to the Registrar of Deaths."* [9]

Dr Oddie issued burial orders for all the victims; after the burial in the mass grave, the Westminster Registrar would have received the notification of the date and place of disposal from the incumbent of St Mary's Church, Cardington. The coroner subsequently held an inquest and although all records have been destroyed, reports indicate that a verdict of accidental death was recorded.

TUJ furnished further details about the events in London:

> *"Messrs. Mills and Sons, of Paddington, as indicated, supervised the removals from Victoria Station to Westminster Mortuary. A guard was then mounted by the R.A.F., and until the Coroner had given his certificates, nothing further could be done. The mortuary itself had already been transformed by means of purple drapery, palms, candlesticks, etc., into something resembling a Chapelle Ardente, but the premises of course did not lend themselves very easily to being beautified. All day Wednesday and Thursday sad groups of relatives arrived at the mortuary to inspect the effects in the small boxes into which they had been placed, and brought from France, and by this means twenty further bodies were identified. (As five were identified in France ... twenty-six in all were positively identified.) By eleven o'clock on Thursday night, therefore, twenty-six breast plates had been engraved and affixed, and all was in readiness for the removal to Westminster Hall. This took place with the greatest dispatch, everyone concerned being a little more accustomed by then to handling such a large*

number of coffins. It must be remembered that each
coffin was metallic lined, and the troops forming the
bearer parties were quite unaccustomed to such work;
the fact that everything was carried out so smoothly,
reflects no little credit upon the organisation of Messrs.
Mills and Sons, who had the supervision of every detail
at every stage.'[10]

While the *Sunday Pictorial* noted that all the nameplates were cut
from the same metal as that used in the construction of the R101,
not all the bodies were identified. However:

'"The Union Jacks which were supplied by the Office of
Works when the coffins were removed to Westminster
Hall were actually buried on the coffins, and therefore
no member of the public was aware that some coffins
bore plates and some did not."' [11]

THE LYING IN STATE

On the advice of his ministers, King George V gave permission
for Westminster Hall to be used for a lying in state. It was an
unprecedented occasion as only deceased royalty had been
privileged to receive such treatment.[12] The coffins covered in
Union Jack flags were transferred from the mortuary very early
on the morning of Friday 10 October and positioned head-to-toe
on purple-draped biers in the hall. A bank of flowers was created
at either end of the coffins, with large arrangements positioned
against tripod stands. A total of 89, 272 people attended the Lying
in State, with queues stretching down Millbank to Vauxhall
Bridge. During the lying in state, a service was held in St Paul's
Cathedral; the king was represented by the Prince of Wales.
Services were also held at the same time in other churches in
areas such as Portsmouth, Liverpool, Peterborough and Sheffield,
while a Mass was celebrated in Westminster Cathedral.

THE JOURNEY FROM LONDON
TO CARDINGTON

On Saturday, 24 army wagons in two equal sections were used
to transport the coffins. Forming in the courtyard outside St
Stephen's Porch, the procession made its way into Parliament
Square, to Whitehall and Charing Cross, then along Aldwych,
Kingsway and Southampton Row to Euston station. The coffins
were placed onto a train on platform number 6 for the 50-mile

▲ The lying in state at Westminster Hall

◀ Some of the large floral tributes displayed inside Westminster Hall

▶ The procession on Kingsway. Two coffins were each conveyed on 24 horse-drawn wagons

▲ A coffin being removed by six RAF bearers from the train at St John's station, Bedford

journey to St John's station, Bedford, on the line to Bletchley. The train left Euston at 12.30 and arrived at 2 p.m.

THE BURIAL

The Times announced that all the relatives of the dead had given their consent to burial in one grave at St Mary's churchyard at Cardington, a short distance from the gates of the Royal Airship Works where the R101 was constructed and commenced its ill-fated trip.

When the train arrived at St John's station, each coffin was removed by a bearing party comprising six RAF personnel, and then placed on 48 army wagons for the three-mile journey to Cardington. Including those walking, the procession was reported to be 1½ miles in length; an estimated 75,000 people were present along the route. It was twilight by the time the coffins started arriving at the churchyard. Pathé Superior Gazette filmed the slow-moving procession and also the burial.

▶ Part of the procession of 48 army wagons to Cardington

▸ Coffins being transferred from the vehicles to the grave.

The coffins were taken through the main gate while a temporary staircase was erected over the wall of the churchyard to allow pedestrian access. From examining photographs and film footage of the scene, a limit appears to have been in placed on the number present for the burial. Around 500 policemen were present, along with 100 ambulance personnel.

Concerning the preparation of the grave, *TUJ* records that:

> *"'Mr Burgess, of Kensal Green Cemetery, at the request of Messrs Mills and Sons, very kindly loaned one of his foreman gravediggers to go down and superintend the digging and building of this tremendous grave. Local labour was employed for the actual digging, and Mr W T Hunt and Mr Evans, the foreman gravedigger in question, made superhuman efforts to have everything completed in time. Artificial turf and about 500 real flowers were used for the decoration of the grave, and many cartloads of earth had to be taken away to make room for the 48 coffins.'"* [13]

From examining photographs, the estimated size of the grave was 29 ft long and 23 ft wide (it was said to be nearly as big as the lounge onboard the R101). A ramp leading downwards to facilitate access for the bearers was constructed at one end of the grave. The coffins were positioned in six rows of eight, but the location of each one was not recorded. Each coffin was covered by a flag upon which rose petals had been scattered. It was not possible to determine whether the remains in each coffin had been identified or not. Contemporary accounts of the burial reveal that the sides of the grave were hung with artificial grass and there were scattered bronze chrysanthemums, carnations and gladioli. Over 3,000 wreaths were displayed near the grave.

▸ A view of the coffins positioned in the grave

◂ The burial service

▸ Some of the floral tributes. Note the tribute made in the shape of the R101

▶ The memorial over the grave in Cardington the churchyard

Four services were conducted by ministers from Anglican, Presbyterian, Wesleyan and Roman Catholic denominations. After the blessing given by the Bishop of St Albans, the silence was broken by a firing party, and the Last Post was answered by the Reveille.

During one of the services the following words were spoken: 'We which are alive and remain shall be caught up together with them in the clouds.'

AN AFTERWORD

An altar tomb memorial, designed by Richardson and Gill, was erected over the grave at Cardington in September 1931. The inscription reads: '"Here lies the bodies of 48 officers and men who perished in HM Airship R101 at Beauvais, France on 5th October 1930."'

Public subscriptions raised £1,000 to complete the work, with the surplus money collected contributing to a fund for the survivors.

Bernard Lane continued to work as a funeral director in Paris until the German invasion of France. In 1940 he moved to London and worked with George Lear as a freelance embalmer before serving with the Royal Artillery in England. In March 1946, Lane was appointed Technical Consultant on matters of embalming, cemetery planning and grave registration by the American Graves Commission in France, and in August 1947 he was mentioned in dispatches for his services in Grave Registration and Enquiries in North Africa and Italy during the last world war. Returning to the family business, Lane embalmed General Prince Arsene Karageorgevitch of Yugoslavia in 1939 and Cardinal Suhard, Archbishop of Paris, ten years later. He died in 1972.

1. See Barclay R (2008) *We're Down Lads. The Tragedy of the Airship R101* Cardington: St Mary's Church; Coates T (ed) (2001) *Tragic Journeys* London: The Stationery Office; Masefield PR (1982) *To Ride the Storm. The Story of the Airship R101* London; William Kimber, Walmsley N Le N (2001) *R101: A Pictorial History* Stroud: Sutton Publishing

2. Parsons B (2005) 'Halford Mills: Funeral Reformer and Pioneer of Embalming' *FSJ* June pp64–72

3. Lane B J (1930) 'R101' *TUJ* October p351

4. 'British Funeral Directors called for the World's Greatest Air Tragedy' (1930) *BUA Monthly* November p115

5. ibid.

6. ibid.

7. Lane B J (1930) 'R101' *TUJ* October p351

8. *The Times* 8 October 1930 p14

9. Oddie S I (1941) *Inquest* London: Hutchinson p218

10. 'More about the R101' (1930) *TUJ* November p370

11. *Sunday Pictorial* 12 October 1930

12. Pond C (2002) *Lying in State* House of Commons Library (SN/PC/1735)

13. 'More about the R101' (1930) *TUJ* November p370.

CHAPTER 11
FUNERAL SERVICE DURING WWII

▲ Recovering victims of the London Blitz

FUNERAL DIRECTORS PREPARE FOR WAR

The possibility of conflict with Germany was anticipated by the mid-1930s, and military and civilian preparations were in place by the time war was declared in 1939. Although a reserved occupation, funeral directors nevertheless recognised that their services would be affected and at the 1938 NAFD conference, the national president, John Jobson, appointed five members to form a Special (War Emergency) Committee to deal with specific war-related issues.[1]

On a local level, one way funeral directors could assist was by becoming involved with the Air Raid Precautions (ARP) department established in each area. In October 1938, the ARP department at Kensington Borough Council approached funeral directors about training them in the decontamination of a gassed body, along with identification and storage until burial.[2] However, the most important development in respect of funerals was the Ministry of Health's circular 1779 'Civilian Death due to War Operations'. Sent to all local authority clerks, it outlined their responsibility for providing mortuary accommodation along with staffing and administration, arranging transport from the place of death to the mortuary and for burial.

Several comments can be made about this circular. First, although it was assumed that most bodies would be claimed by relatives who would then arrange and pay for a funeral, where this did not take place the local authority would be obliged to carry out a burial. Secondly, authorities were required to assess available burial space, and were also empowered to use burial grounds not under their control, including proprietary cemeteries and churchyards.[3] Thirdly, there was no mention of cremation: with only 3.5 per cent of deaths followed by cremation at the 54 crematoria operating in 1939, it was not the preferred mode of disposal. Furthermore, the National Association of Cemetery and Crematorium Superintendents (NACCS) had already pointed out that crematoria would be impeded if gas and electricity supplies were severed.[4] However, others disagreed and at the eighth joint conference of cemetery and crematorium authorities, Dr A B Williamson of Portsmouth stated: 'There is probably no stronger advocate of cremation than the Medical Officer of Health.'[5]

In addition to NACCS's point, there were objections to cremation by a large proportion of the community (despite, for example, Hull providing the incentive of a free cremation service and St Marylebone Crematorium reducing their fee to £2 2s for war victims), while the cremation of the unidentified and delays due to certification were important factors.[6] Another difficulty was that the cremation of unidentified bodies prevented exhumation for identification at a later date. Lastly, consultation with funeral directors by the local authority was encouraged. Although funeral directors did not receive the circular, it was reprinted in *TUJ* and *TNFD*.[7] The NAFD recommended that for reasons of unsuitability, funeral directors did not offer their premises as mortuary accommodation, that staff in the mortuaries should not be connected with funeral directors and that vehicles used for the recovery of bodies should not be hearses due to the possibility of transferring gas-contaminated bodies.[8] However, as the war progressed, many members gave their time freely as mortuary attendants, while members' premises were commandeered for body storage and funeral vehicles were utilised.

Local authorities opened mortuaries in buildings such as recreation halls and swimming baths, while newly recruited staff were trained in interviewing the bereaved and identification techniques: those at St Pancras mortuary in London, for example, were trained by Sir Bernard Spilsbury, honorary pathologist to the Home Office; Bentley Purchase, the coroner; and a police inspector.[9] The Borough of Twickenham converted two open-air swimming baths and the cemetery chapels into temporary mortuaries. Capable of accommodating 244 bodies, the cost of

▼ A Borough of Camberwell advertisement for mortuary personnel (*TNFD*)

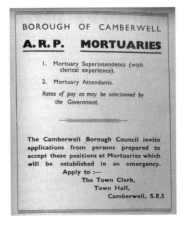

▲ The Coventry and District Funeral Directors' Guild assemble for a photograph during a respirator drill in 1939. Mr A E Pargetter was chairman of the Funeral Directors' ARP Committee (*TNFD*)

racking, water boilers and other fittings was £990.[10] Bodies were to be labelled with a metal or Bakelite disc with a serial number; where appropriate, photographs were also to be taken.[11]

In Croydon, the ARP department estimated that 140 men would be required for their 'collecting party' to serve four mortuaries capable of holding 500 bodies, and the local association contacted the council to offer assistance: the Coventry Guild of Funeral Directors did the same.[12] The York Association took part in a mock air raid involving twelve 'deaths'. The London-based wholesale suppliers Dottridge Bros was of sufficient size to establish their own Local Defence Volunteer or 'Home Guard' unit.

SEPTEMBER 1939: WAR IS DECLARED

Following the declaration of war, funeral directors were immediately affected by restrictions. Firstly, although classified as a reserved occupation in July 1918, funeral directors (including assistant directors, branch managers, foreman coffin makers and coffin makers) aged 30 years of age and under were called up.[13] Hearse drivers were later added to the list. Noticeably absent from this list were embalmers, probably as there were very few full-time practitioners and comparatively few treatments were carried out. Despite deputations on a number of occasions to government ministers from the NAFD's Special (War) Emergency Committee, the age for call-up was increased to 35 from October 1941.[14] Secondly, petrol for hearses, following cars and removal vehicles was rationed.[15] The same committee met with the Ministry of Mines (Petroleum Department) in September 1939, which confirmed that applications for supplementary supplies would be considered.[16] However, as the war progressed, attempts were made to change working practices in an effort to conserve supplies. The suggestion of motor hearses only carrying out funerals within a fixed radius was first made in 1940, but soon shelved.[17] However, by June 1942, the Ministry of War Transport stated that if a coffin had to be transported over 30 miles 'in ordinary circumstances', this had to be by rail, except if many changes were involved.[18] The use of vehicles for long-distance funerals was discouraged (as will be seen, this was only one initiative concerning funeral transport). Thirdly, the rationing of tyres was introduced in 1942 and supplied only for vehicles essential to the war effort.[19] Lastly, restrictions on timber were introduced at the end of September 1939: any person purchasing timber or boxboard up to a value of £20 per month had to have a licence.[20] By December, this was reduced to £5 per month, while the following year the thickness of wood used in coffin construction had been reduced: the sides, ends, tops and bottoms were not to exceed 5/8" thick.[21] Other restrictions affected coffin preparation: the availability of metal fittings for caskets ceased in May 1942, while sandpaper was also rationed.[22]

CONTRIBUTIONS TO THE WAR EFFORT

In addition to the above restrictions, the industry was also involved in a number of initiatives to help the war effort. Towards the end of 1939, the Newcastle, Gateshead and District Local Association suggested eliminating funerals at a 'walking pace' so that vehicles could do at least two funerals per day.[23] However, a more radical plan was to stagger funerals and, with the encouragement of the Ministry of War Transport, a pilot scheme commenced at Bolton in 1942.[24] Traditionally, most funerals had taken place between 2 p.m. and 3 p.m. By expanding this to 10 a.m. and 3 p.m., a reduction in the use of hearses from 21 to 10 was recorded. A central office was used to co-ordinate vehicles and arrange times with funeral directors. Furthermore, processions where mourners walked in front of or behind the hearse were not allowed, as a minimum speed of 15mph was introduced; mourners were encouraged to meet funerals at the church; and processions were to take the most direct route to a cemetery or crematorium.[25] The staggering scheme was also proposed in Portsmouth and Glasgow.[26]

As has been previously stated, the interwar years were a period of transition from the use of horse-drawn to motor hearses: the war did not help with retaining horses as feed was subject to rationing. The Cemeteries Association was also concerned that horses would be difficult to manage in case of gunfire or explosions, and suggested that their use be curtailed.[27]

Other war-effort suggestions included the Southampton local association proposing to reduce brass or copper coffin handles from four to two pairs, while W H Painter of Birmingham encouraged funeral directors with coffin plates containing mistakes and tarnished metal fittings to send them for scrap.[28] Railings from graves and the boundaries of cemeteries were removed for the same reason.[29]

A degree of flexibility was introduced as cemetery authorities changed their hours of burials, while Registrars of Deaths also extended their attendance times.[30] Funeral directors were encouraged to become involved in fire watching.[31]

The NAFD had to modify its annual gatherings to save on long-distance travel for members; the 1940 conference was scaled down from three days in Aberystwyth to one day in Manchester, and a one-day event was held throughout the remaining war years except 1944, when the conference was cancelled.[32]

More general restrictions also impacted on funeral directors. If an air raid sounded when a funeral was on its way to a cemetery or crematorium, the procession had to leave the street to keep them clear for civil defence vehicles.[33] Gas masks had to be carried during funerals, while vehicle lights had to be modified to expose the beam only through an aperture the size of a half-penny.[34] Speed limits were enforced in the blackouts with offenders being fined by the magistrates.[35] Signs and traffic lights

▼ Sandbags protect the premises of Thomas Ebbutt funeral directors in Croydon

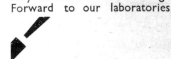
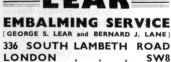

WAR

... has presented your business with many unusual difficulties

Supplies, personnel, and equipment each present their special problem

Our contacts with the Ministry of Supply and the Ministry of Transport ensure that within the limits of necessary wartime economy we have every hope of fulfilling requirements

Our productive capacity is at your service

1835 · 1940

Dottridge Brothers Ltd.

DORSET WORKS, EAST ROAD, CITY ROAD, LONDON, N.1

TELEGRAMS: "FUNERAL, LONDON" · TELEPHONE : CLERKENWELL 1170 (PRIVATE BRANCH EXCHANGE)

◄ Suppliers to the industry were affected by the war. The coachbuilder Thomas Startin ceased making hearses and limousines for most of the war, while George Lear appealed for glass embalming-fluid bottles to be returned for reuse. Supplies from Dottridge Bros were curtailed and their premises in London bombed on fifteen occasions (*TNFD*)

countrywide were removed too, so, to help with navigation, the NAFD established a scheme whereby a local funeral director would pilot a visiting funeral through a town.[36] All premises had to be blacked out and, in some cases, protected by sandbags.

FUNERALS OF THE VICTIMS

Despite the challenges of war, funerals carried on, and, as in WWI, funeral directors had also to cope with deaths euphemistically described as being due to 'war operations'.[37] It is estimated that there were over 67,000 civilian victims during WWII. As mentioned above, families could claim bodies of relatives and arrange for their burial. However, under Circular 1779, the local authority had an obligation to bury unidentified and/ or unclaimed victims. After a discussion between the NAFD's national secretary and the Ministry of Health in October 1940, it was recorded that:

> '*Where [unidentified] casualties are numerous, the local authorities have provided for disposal in canvas wrappings or cardboard coffins, in shallow trenches, so as to facilitate any exhumation if it becomes necessary at a later date.*'[38]

Local authorities were to approach the Ministry of Health with the number of shrouds they wished to purchase; the minutes of Bermondsey Borough Council indicate that the ministry could obtain them for 3s 9d each.[39] St Marylebone Borough Council purchased 750 pairs of sheets for shrouds from Peter Robinson in Oxford Street at 2s 4½d per pair.[40] The borough of Twickenham ordered 250 fabric-lined rubbers sheets, and also hessian bags for personal belongings.[41] However, both funeral directors and cemetery superintendents were scornful of the use of anything except a coffin. The editor of *TUJ* commented that:

> '... burial in sheets is a lamentable mistake on the part of the Ministry and likened such burials to a pauper's funeral, while NACCS objected on the grounds of the difficulties attached to disinterring uncoffined bodies.'[42]

In Coventry, rumours that cardboard coffins were to be used was dismissed by the cemetery superintendent, P W H Conn, who was also given the title of 'Director of Civilian Dead in War Time'.[43] He stated, 'If a calamity occurred in Coventry, everyone would receive a decent coffin to rest in.' He was also anxious that use be made of the newly opened Canley Crematorium.[44] This was aided by an amendment made to the Cremation Regulations 1930 to dispense with the need for two medical certificates if death had been registered as being due to war operations.[45]

Although cemetery superintendents and funeral directors had contrasting reasons to object to shroud burials, they did, however, achieve a measure of success in maintaining the use of coffins. A Ministry of Health circular (No 2192) issued towards the end of 1940 stated that where bodies were not claimed and the local authority arranged a funeral, it should take place in a 'seemly and dignified' manner, and that, 'Burial of civilians should be regarded as no less honourable than burial of a soldier by his comrades.'[46] Where a severe strain was placed on a local authority, the '... multiplicity of funerals might in certain conditions be understandable.'[47] The superintendent of a Glasgow cemetery found that some people who resented the idea of the possibility of being buried in a communal grave were reassured that it was not a pauper's grave.[48] Under the Defence Regulations, a payment of £7 10s would be made by local authorities for the burial of a civilian victim of war. However, it was only intended to be a contribution towards funeral costs.

As already mentioned, few bodies were embalmed at this time, although both the BIE and BES were actively encouraging sanitary education among funeral directors through classes and articles in the trade journals. George Lear was working as a full-time trade embalmer and in 1940 reported that embalming was ideal where wartime conditions delayed the burial.[49] He suggested that one reason to carry out embalming was where an interment could not take place due to a time bomb falling in a cemetery, as had happened recently. He also noted that if arterial injection was not possible, derma-surgery could be used. Embalming, however, did have its place: a team of American embalmers based at Brookwood Cemetery in Surrey treated the bodies of US servicemen that were to be repatriated at a later date.[50]

FUNERALS OF THE VICTIMS: COVENTRY

Apart from London, the worst city to be devastated in the Blitz was Coventry. On 14–15 November 1940, a 10-hour attack devastated 100 acres of the city centre: 554 were killed and, according to an account by Kenneth Turner who dealt with the identification of the bodies, even the mortuary was bombed. Mortuary staff, '... were engaged in fitting bits and pieces together to make whole bodies.'[51]

An account of preparations for the mass funeral written by the secretary of the Coventry and District Funeral Directors' Guild, A J Maton, appeared in *TNFD*.[52] He was summoned to a meeting of the town clerk and the medical officer of health on Saturday 16 October, where a decision had already been made for a mass burial using shrouds, as it was not considered possible to obtain 500 coffins. However, the following day, the City's Emergency Committee requested 400–500 coffins to be supplied by Tuesday afternoon. The town clerk's messenger was despatched to visit all

▼ A view of the embalming theatre adjacent to the American military section at Brookwood Cemetery *(TNFD)*

▼ The cathedral church of St Michael's Coventry was bombed on 14 November 1940. In 1947, a group of Dutch funeral directors visited England and were shown around Coventry by members of the local Guild *(TNFD)*

Coventry funeral directors with the instruction to start making coffins and to continue until otherwise ordered. However, many firms had suffered bomb damage, whilst there was also a partial absence of electricity. The Birmingham Guild of Funeral Directors was contacted and W H Painter agreed for his members to provide 250–300 coffins. Through George Jennings in Wolverhampton, funeral directors in the area supplied a further 100, while other associations also offered coffins and support.

The funeral for 172 victims took place on Wednesday 20 November in the extension to Coventry's London Road Cemetery. The coffins were transported to the cemetery on a lorry the night before and placed three deep in two long, parallel trenches that had been dug by soldiers and labourers, assisted by a mechanical digger. The next morning, mourners assembled at the gates and then walked behind the mayor, the bishop of Coventry, clergy and other officials to the grave where a joint committal service was held. A second mass burial was held a few days later.[53]

FUNERALS OF THE VICTIMS: HOLBORN

Bounded by the Cities of Westminster to the west and London to the east, the Metropolitan Borough of Holborn was one of the smallest local authority areas in London, but was as heavily bombed as its neighbours.[54] According to the Borough's record of air raid incidents, 432 civilians were killed between 1940 and 1945. In addition, a number were 'presumed dead'.[55] The Borough's mortuary was located in the basement of a building at Stukeley Street, near Covent Garden.[56] Heeding the Ministry

▼ The church of St Alban the Martyr, Holborn –
before and after the bombing

of Health's circular 1779 and mindful that the borough did not
have its own cemetery, contact was made with other London
authorities for burial space; the Borough of Wandsworth offered
its newly acquired (but never used) Tolworth Cemetery. However,
trench graves were prepared at Putney Vale and Wandsworth.
The worst incident in Holborn was on 8 March 1945 when a V2
rocket killed 135. Burial of six unidentified victims took place in
St Alban the Martyr, Holborn section at Brookwood Cemetery.[57]
A. France & Sons arranged the funerals of a number of civilian
deaths from this and other incidents: entries in the ledgers are
headed 'WO' – War Operations.

FUNERALS OF THE VICTIMS: KINGSTON

The firm of Frederick W Paine carried out many funerals of
air raid victims in the Surbiton and Kingston area; a target for
the bombing being the Hawker Aircraft works on the Thames
at nearby Ham. Run by Frederick Paine, who died just before
the end of the war, and with 11 branches in Middlesex and
north Surrey, it was the largest firm of funeral directors in
Britain. Their records reveal that 254 funerals were carried out
in October 1940; 63 of these were described in the registers as
'Air Raid Casualty'. All funerals used motor vehicles, and coffins
were received from or transported to distant locations such as
Bournemouth, Manchester and Dumfries using the railway.
One feature highlighted by Paine's registers is the number of
multiple funerals: in October 1940, nine were recorded. Five
members of the same family were killed in Hampton at the end
of September 1940; all were buried in the same grave during a
funeral conducted by Frederick Paine and his brother-in-law; the
ceremony required 20 bearers.[58] A direct hit on a keeper's house
at Chessington Zoo resulted in the death of two family members.

The firm also looked after the funeral of a victim of the Balham underground station bombing, when 64 of 600 shelterers were buried alive on 15 October 1940 after a bomb hit the main road above, resulting in water and sludge pouring into the tunnels.

THE IMPACT OF WAR ON FUNERALS

As was the case in WWI, the years of conflict brought challenges for funeral directors, whilst also changing the course of funerals in wider terms. For example, with the payment of £7 10s towards the burial of civilians due to war operations, it was the first time the government had made a contribution towards funeral expenses. The issue of high funeral costs, alleged exploitation by funeral directors and the poor paying for burial insurance to prevent a pauper's funeral were reoccurring themes during the interwar period. A suggested remedy was the municipalisation of funeral directors, and this had been tabled on a number of occasions.[59] In 1938, Sir Arnold Wilson and Professor Herman Levy's detailed study entitled *Burial Reform and Funeral Costs* was published, which contained recommendations for an inquiry into funeral charges and that death benefit be part of the National Health Insurance Acts.

▲ Preparations for burial of the 31 children and one teacher who died when Sandhurst Road School at Catford in south-east London was bombed in January 1943. The white coffins were kept overnight in the cemetery chapel and a communal service held at the trench grave the next morning. Reporting restrictions were lifted for this tragedy, hence this photograph *(Getty Archive)*

▼ ▲ ▶ Censorship rules did not permit the British press to show bombed buildings. These three views of the bombing at London Wall, Moorgate station and St Giles Church, Cripplegate were part of a set of postcards passed by the censor in 1944

◄ Ingall, Parsons, Clive & Co
'Victory' *(TNFD)*

▼ 1944 Lancefield
Coachworks *(TNFD)*

The onset of WWII did not stem the momentum for reform. In July 1942, accusations that funeral directors were exploiting the families of air raid victims in Weston-Super-Mare caused comment in Parliament.[60] However, they proved to be unfounded. The Fabian Society issued a report in 1943 that recommended the introduction of a local authority funeral service, as did a document by The Social Security League the following year.[61] However, midway through the war, William Beveridge published their report *Social Insurance and Allied Services*.[62] Framing what would become the 'Welfare State', the 1942 report recommended payment of a £20 funeral grant. This was eventually introduced in June 1949 in conjunction with the Central Price Regulation Committee.[63] The government's contribution to the burial of civilians was, however, less than half of the 1949 full grant: even though increased to £10 in 1944, it was far lower than the average cost of a funeral. An assessment of Frederick W Paine's funeral records shows that the average charge for a funeral in 1940 (without disbursements) was in the region of £18. It was 1948 before the Death Grant was finally introduced.

That the war was a route towards change and progressive opportunities is clear from *TUJ* and *TNFD*. In an article entitled, 'Age-old customs gone in a couple of years' published in 1943, the writer summarised changes attributable to the war, including the pooling of resources, the staggering of funeral hours, no 'walking' funerals and that cemeteries were giving credit for fees.[64]

As early as 1941, the NAFD conference was used as a platform to encourage change. In that year, the national president, W H Painter, put forward three suggestions: development of the association; resume education work; and rekindle the objective of state registration.[65] The national secretary, George Hotter, made similar pleas during the war years, whilst Florence Hurry, the assistant secretary, presented a paper on funeral service education.[66]

It was the combination of factors occurring during the first half of the twentieth century that presented opportunities for funeral service to develop, in addition to reforming the manner in which funerals were carried out. Changes included greater responsibility for custody of the body as death increasingly occurred in hospital, especially following the establishment of the National Health Service in 1948.[67] Funeral directors responded by providing chapels of rest while sanitary education was given a new impetus.[68]

Funeral directors were also encouraged to assess the strength of their business and future opportunities. Included within the pages of *TUJ* in the month the war ended was a guidance document entitled 'Planning Ahead'.[69] Transport was changing with the motor hearse triumphing; in 1951, it was reported that the last horse-drawn funeral had taken place in London.[70]

The immediate post-war years also gave the promotion of cremation a boost as the Cremation Society of Great Britain encouraged local authorities to build much-needed housing rather than open cemeteries: its president, Lord Horder, pressed for 'playing fields not cemeteries.'[71] Although five crematoria opened during the war years, by 1945 the preference for cremation had only moved from 3.5 per cent to 7.8 per cent.[72] However, 10 years later, with 24 new crematoria opened, one quarter of deaths would be followed by cremation.

In comparing the challenges facing funeral service during WWII with the previous period of conflict, not only did restrictions on labour and materials have a similar impact, but additional difficulties were caused by bombing, together with the need to bury civilian victims. Nevertheless, funeral directors worked together to ensure the dead reached their resting place, whilst helping to maintain the British spirit of defiance, stoicism and self-sacrifice during a time of national emergency: one that was also a prelude to a future of change not only for the nation, but also funeral service.

▼ Victory: The Home Guard marches past F A Albin's premises in Bermondsey on VE day in May 1945 (Courtesy of Barry Albin)

1. 'The Discussion of the National Council Report: War Emergency' (1939) *TNFD* August pp46–49

2. 'Editorial ARP Defects' (1938) *TUFDJ* October p369 and p371. See also 'You and ARP' (1939) *TNFD* January p263, and 'Discussion on the Annual report. War Emergency' (1939) *TUFDJ* July pp245–247, and 'ARP Burial Section. How Uxbridge Started' (1938) *TUFDJ* December p443

3. The superintendent at St Marylebone Cemetery reported that 15,214 common graves and 63,826 private graves were available and 2,500 could be located. See St Marylebone Borough Council minutes 27 April 1939 p478

4. Rugg J (2004) 'Managing "Civilian Death due to War Operation": Yorkshire Experiences during World War II' *Twentieth Century British History* Vol 15 No 2 p160 and also Rugg J (2005) 'Managing "Civilian Death due to War Operations": The Lessons of World War II' *The Journal of the Institute of Cemetery and Crematorium Management* Vol 73 No 1 pp40–42

5. 'The Problems of the Disposal of the Dead after Air Raids' (1939) *TUFDJ* October p357–361. See also 'Portsmouth's Procedure' (1939) *TNFD* September p104

6. 'Free Cremation' (1939) *TNFD* September p134 and St Marylebone Borough Council minutes 20 June 1940 p447

7. 'The Ministry of Health's Circular No 1779' (1939) *TNFD* May pp414–415 and 'Civilian Dead During War' (1939) *TUFDJ* April pp121–122

8. 'Funeral Directors and War Service' (1939) *TNFD* May pp410–411. See also Hebblewaite AS (1939) 'ARP' *TNFD* June pp479–482 for ARP arrangements in Sunderland

9. Hart V & Marshall L (1983) *Wartime Camden* London: Camden. Sir Bernard Spilsbury, Bentley Purchase and Leverton's funeral directors were involved in Operation Mincemeat: see Macintyre B (2010) *Operation Mincemeat. The True Spy Story that Changed the Course of World War II* London: Bloomsbury

10. Borough of Twickenham Council minutes 19 June 1939 pp646–647 and 21 July 1939 p823. See also Barnfield P (2007) *When the Bombs Fell. Twickenham, Teddington and the Hamptons under Aerial Bombardment during the Second World War* Twickenham: Borough of Twickenham Local History Society

11. *War Deaths: Emergency Mortuary and Burial Arrangements* HMSO: Ministry of Health paragraphs 22 and 36

12. 'Air Raid Victims' (1939) *TNFD* September p104, and 'Coventry Directors and ARP Work' (1939) *TNFD* September p114 and Maton A (1939) 'Coventry Funeral Directors in ARP Work' *TUFDJ* September p335

13. 'The April Meeting of the National Council' (1939) *TNFD* May pp426–428. See also 'The Report of the National Council' (1939) August p42; 'Reserved Occupations' (1941) *TNFD* February p242; 'The War and the Funeral Trade' (1939) *TUFDJ* October pp351–352; 'Reserved Occupations' (1940) *TNFD* August p64; 'Reserved Occupations' (1941) *TNFD* April p310. See also 'Reserved Occupations' (1941) *TNFD* September (loose page in publication); 'The New Schedule' (1941) *TNFD* May p344; 'Official Notice: Reserved Occupations' (1941) *TUFDJ* July p163. See also 'Reserved Occupations' (1941) *TUFDJ* September pp239–242 and The National Council's Record Elucidation of Many Matters and Discussion Thereon' (1942) *TNFD* August p50. Lord Terrington (1889–1961) wrote a preface for the NAFD's *Study Course in Funeral Management* published in 1948

14. Call-up was finally suspended in April 1944. See 'Man-Power-Latest on Deferments' (1944) *TNFD* April p385

15. 'The War and the Funeral Trade' (1939) *TUFDJ* October pp351–352

16. 'War Information: Petrol Supplies' (1939) *TNFD* October p147. See also 'What the NAFD has been doing Pre-war and War-Activities' (1939) *TNFD* October p157

17. 'Restricted Radius for Motor Hearses' (1940) *TNFD* April p348

18. 'Petrol for Hearses' (1942) *TNFD* July p4. In Scotland this was 20 miles. See 'The Motor-Hearse Limit' (1943) *TNFD* June p462 and 'The Thirty Mile Limit' (1943) *TNFD* August p60. See also 'The 30-mile Limit' (1943) *TNFD* November pp180–181

19. 'Tyre Supplies' (1942) *TNFD* May p344

20. 'Control of Timber Supplies' (1939) *TUFDJ* October p353

21. 'Ministry of Supply Timber Control' (1940) *TUFDJ* January p20, and 'Timber and Mouldings' (1940) *TNFD* August p35. See also 'The Timber Shortage' (1941) *TUFDJ* October p273

22. 'The Coffin Furniture Manufacturers' Association' (1942) *TUFDJ* May p94 and *TNFD* May (1942) p245, and 'Your Sandpaper Requirements' (1942) *TNFD* May p346. Bunny France recalls that zinc for lining coffins could only be

obtained by seeking written permission from the Ministry of Supply. With no repatriations taking place during the war years, bodies were placed in zinc-lined coffins and retained in catacombs, such as at St Mary's Cemetery in Kensal Green, to await eventual transportation.

23. 'Notes' (1939) *TUFDJ* November p381 and 'No More Walking Funerals' (1939) *TNFD* November p181 and 'Plea for Simpler Funerals' (1940) *TNFD* September p79. See also 'Minimum Speed of 15mph for Funerals' (1944) *TUFDJ* June p165

24. 'Staggered Funerals to Save Petrol' (1942) *TUFDJ* June p121. See also 'Editorial: Staggering Funeral Times' (1942) *TUFDJ* June p125. See also 'Pooling Resources' (1943) *TUFDJ* March p79. See also 'The Yorkshire Quarterly: Petrol and "Staggered" Funerals' (1942) *TNFD* May pp354–356 and 'Funerals Pools' (1943) *TNFD* March pp325–326, 'Reports from Nor'-West' (1943) *TNFD* June p450

25. 'Minimum Speed of Funeral Vehicles' (1944) *TNFD* October p126, and 'Walking Funerals' (1942) *TNFD* January p247. See also 'Walking Funerals' (1943) *TNFD* February p294 and 'Walking Funerals' (1943) *TNFD* June p467

26. 'News from the Provinces: Portsmouth' (1942) *TNFD* March p292 and September p126. See also 'Glasgow's Pooling Scheme' (1942) *TUFDJ* October p237

27. 'Use of Horses in Wartime' (1939) *TNFD* October p162

28. 'Non-Ferrous Metals' (1942) *TNFD* November p165

29. 'Collection of Railings' and 'Railings of Tombs' (1942) *TUFDJ* February p50. See also 'Mobilising the Churchyards' (1942) *TUFDJ* June p133 and 'Grave Railings for Salvage' (1941) *TUFDJ* October p274

30. 'London Official Notices' (1939) *TNFD* November pp186–188

31. 'Fire Watchers' (1941) *TNFD* February p243 and 'Compulsory Fire Watch at Last' (1941) *TNFD* February p257. Bunney France, of A. France in Holborn, recalls taking his turn on the roof of the firm's premises in Lamb's Conduit Street, where a stirrup pump and sand were kept for putting out fires.

32. 'Cancellation of the Aberystwyth Conference' (1940) *TNFD* June p413. See also 'Notes' (1940) *TUFDJ* June p147

33. 'The War and the Funeral Trade' (1939) *TUFDJ* October p352. See also 'Air Raid Warnings' (1939) *TNFD* October p150, and also 'Securing of Horses (Defence) Order, 1940' (1940) *TNFD* December p174

34. 'Your Car Lights' (1940) *TNFD* November p144

35. 'In a Hurry' and 'Hearses did 30mph' (1940) *TNFD* April p349

36. 'Piloting Through Towns' (1941) *TNFD* January p208

37. 'The War and the Funeral Trade' (1939) *TUFDJ* October pp351–352

38. 'Civilian Deaths as a Result of Air Raids' (1940) *TNFD* December p186

39. Bermondsey Borough Council minutes 10 July 1939 p164

40. St Marylebone Borough Council minutes 27 April 1939 p713

41. Borough of Twickenham Council minutes 19 June 1939 p646 and 21 July 1939 p823

42. 'Editorial: Wartime Burials' (1939) *TUFDJ* October p367

43. 'North Western Area Federation Quarterly Meeting (1939) *TNFD* November pp198–199

44. 'Coventry's New Crematorium and Cemetery; Fifty-eighth in Great Britain' (1943) *TUFDJ* March p79

45. 'Defence Regulations' (1940) *TUFDJ* August p218. See also 'Cremation in the Case of Deaths Occurring to Civilians or Members of His Majesty's Forces in Consequences of War Operations' (1940) *TNFD* August p63; Piggott H (1939) 'Cremation and A R P' *Pharos* Vol 5 No 3 pp7–8; 'Cremation of Civilian Dead in Wartime. The New Regulations Explained' (1940) *Pharos* Vol 6 no 6 p2

46. 'At Last' (1941) *TUFDJ* January p12

47. 'Burial of Air Raid Victims' (1940) *TUFDJ* November p283. See also 'Editorial: Specialised Service' (1940) *TUFDJ* November pp295–296. See also 'Civilian Deaths Due to Air Raids' (1941) *TUFDJ* July pp173–174 and 'Civilian War Deaths Due to Air Raids' (1941) *TNFD* September p86 and 'Funerals after Raids' (1941) *TNFD* December p185

48. Dalglish T D (1944) 'After the 1941 Blitz: Cemeteries Difficulties Overcome' *TUFDJ* June pp168–169

49. 'Embalming Air Raid Victims' (1940) *TUFDJ* October p259–261. See also 'Preparing the body for Post-Mortem Derma-Surgery' (1940) *TNFD* September p84

50. 'Burial of American Army Casualties. Mr Alfred Morrison Visits Brookwood' (1943) *TUFDJ* May p119. See also 'Brookwood American Cemetery. An Impression' (1943) *TUFDJ* September p231 and 'United States War Dead in England' (1943) *TNFD* January pp256–257. For further references to embalming during WW2 see *Tovey Bros: A 150 year history in Newport* (2010) Newport: Tovey Bros pp34–36

51. Lewis T (1990) *Moonlight Sonata: The Coventry Blitz 14/15 November 1940* Coventry p169

52. Maton A J (1941) 'Coventry in the Blitz' *TNFD* February p249 and *TUFDJ* January (1941) p49. See also Jalland P (2010) *Death in War and Peace. The History of Loss and Grief in England 1914–1970* Oxford: Oxford University Press pp121–139

53. See Harkin T (2010) *Coventry 14th/15th November 1940. Casualties, Awards and Accounts* Coventry: War Memorial Park Publications and Lewis T (1990) *Moonlight Sonata: The Coventry Blitz 14/15 November 1940* Coventry

54. For a history of domestic wartime activities in Holborn's neighbouring borough see Newberry C A (2006) *Wartime St Pancras: A London Borough Defends Itself* London: Camden History Society. See also the poem 'Disposal of Civilian Dead' by Nick Durrant in Mannix A and Lewin E Eds (2010) *Postcards from Leather Lane* Camden: London Borough of Camden/Spread the Word

55. The procedure to establish 'presumed dead' status was simplified by the Defence Regulations 30A. If the coroner was satisfied that a person was believed to have been killed by enemy action and that the body could not be identified or found, it would be registered as such by the Registrar of Deaths. See St Marylebone Borough Council minutes 22 May 1941 p135

56. Bunney France recalls seeing bodies there in the mortuary there, wrapped in hessian sheets not dissimilar to the sketches of people sleeping in shelters by the war artist and sculptor, Henry Moore.

57. Parsons B (2011) 'The Civilian War Grave in St Alban's Burial Ground at Brookwood Cemetery' *Necropolis News* Vol 5 No 4 pp3–8. See also Record of Air Raid Incidents, Metropolitan Borough of Holborn

58. See Borough of Twickenham Register of Civil Deaths due to Enemy Action [L.940.546]

59. 'Municipal Funerals: A Liverpool Scheme' (1916) TUJ June p145 and 'Alderman Taggart on £5 Funerals' (1916) TUJ July p172. See also 'Funeral Reform' (1916) *The Hospital* 17 June Vol LX No 1566 p252. See also 'Funeral Reform' (1916) *The Hospital* 29 July Vol LX No 1573 p404

60. 'Weston-Super-Mare' (1942) *TNFD* September p110. See also 'Touting by Weston Undertakers Alleged' *The Weston Mercury and Somersetshire Herald* 25 July 1942 p5

61. 'Funeral Reform' (1943) *Fabian Quarterly* January pp23–29. Clarke J S (1944) 'Funeral Reform' *TUFDJ* August pp229–230

62. Beveridge W (1942) *Social Insurance and Allied Services* London: HMSO pp65–66. See also 'Funeral Allowances in New Plan' (1942) *TUFDJ* December p297; 'The Beveridge Plan' (1943) *TNFD* January p246; 'Beveridge Funeral Grant' (1943) *TUFDJ* April p89; and 'Government Social Insurance Proposals' (1944) *TUFDJ* October p294

63. 'Funeral Prices Regulated' (1949) *FSJ* June p321

64. 'There's Something to be said for Leadership' (1943) *TNFD* October p133. See also 'From War to Peace' (1944) *TUFDJ* September p249

65. 'The National Council Meet in Harrogate' (1941) *TNFD* November pp148–150

66. Hotter G (1942) 'The NAFD in War Time and After' *TNFD* August pp68–71 and 'Editorially: Trade After the War' (1943) *TNFD* September p101 and Hotter G (1945) 'Post War Activity' *TNFD* April pp367–370. 'A Wider Outlook' (1943) *TUFDJ* August p214. See Hurry F D (1943) 'A Wider Outlook' *TNFD* August pp74–75. See also 'The Educational Policy: A Big Step Forward' (1943) *TNFD* November pp172–173; Long J C (1943) 'Our Educational Policy' *TNFD* December p212–213; 'What is Education? By a Layman' (1944) *TNFD* February pp296–297; 'The Complete Education Policy Outlined' (1945) *TNFD* March pp344–345

67. 'From War to Peace' (1944) *TUFDJ* September p249

68. 'New School of Embalming and Funeral Hygiene' (1942) *TUFDJ* July p193

69. 'Planning Ahead! An Aid to Post-War Development Incorporating Progressive Ideas' (1945) *TUFDJ* May p149

70. Parsons B (2008) 'Transport to Paradise – Part 1: The Horse-Drawn Hearse' *FSJ* September pp97–119

71. Horder Lord (1942) 'Cremation and Post-War Planning' *TUFDJ* June p130

72. For 'The Kensal Green Crematorium' (1939) *TUFDJ* November pp385–387; 'A New Crematorium' (1940) *TNFD* February p280; 'New Cemetery and Crematorium' (1940) *TNFD* June p414; 'Kettering Crematorium' (1940) *TUFDJ* July p195; 'Coventry's New Crematorium and Cemetery; Fifty-eighth in Great Britain' (1943) *TUFDJ* March p79; 'Britain's 58th Crematorium. Bishop of Lewis Dedicates Brighton's New Building' (1941) *TUFDJ* May pp212–122.

The Undertaker at Work: A History in Images

PREMISES

▼ ▲ Few images of undertaking premises exist from the early part
of the twentieth century. However, in April 1916, *TUJ* published
photographs of Henry Sherry's new branch office at Park Road, St
John's Wood. The article quaintly noted: 'There is nothing at all
suggestive of the lugubriousness of the final ends of mortality. Of
course one can imagine that when a client calls the needful articles
of sepulture are not far to seek, but they are not thrown obtrusively
at you. There is not even a shaving on view or patent to the olfactory
sense' *(TUJ)*

▶ ▼ The plan of an unidentified undertaking premises in Liverpool from around 1919 showing the location of the small mortuary, which was, somewhat curiously, accessed through the men's dining room. The remaining premises comprise stables, harness room and covered shed. The second image shows the interior of the chapel of rest with painted bricks, decorative dado and altar. The dark wood coffin with raised lid and 'acorn' type coffin-closing screws is from the higher end of the range

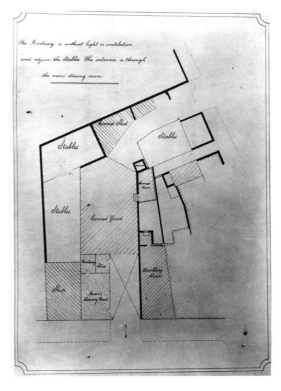

◄ The exterior of James Hawes, funeral furnisher, in Station Road, Manor Park in east London. Much use is made of the window space to display white marble statues of angels and a large cross, in addition to a range of 'Immortelles' or 'French wreaths' for placing on a grave. These were china or porcelain flowers wired to a base and sealed under a glass dome, cross, anchor or other design. Purchased from undertakers, they were very popular in the late nineteenth and early twentieth centuries. Several firms supplied the market: the French Wreath Company had a factory in Paris and a showroom on Tottenham Court Road; Messrs J S Robinson was situated at 270 Bishopsgate in the City of London; Osman & Co was at 157 Commercial Street; and Zobel & Sons on the Euston Road

▼ In the post WWI years, an increasing number of undertakers provided accommodation for the body; the availability of this facility was often included in newspaper advertising and on letterheadings. The term 'private mortuary' was used at first, but 'chapel of rest' was eventually adopted as the word 'mortuary' was linked with the cold, communal space for unclaimed and/or infectious bodies. In contrast, a 'chapel of rest' fostered an image of a sacred and exclusive environment in which the body could remain undisturbed until the funeral. The chapel opened by Thomas Porter & Sons in their premises at Dingle in Liverpool was opened in November 1919 and was in frequent use

◄ One of the austere tiled chapels at the London Necropolis Company's station in Westminster Bridge Road, which was built in 1902. Coffins would rest in this chapel before being placed in a hearse van to be conveyed by rail to Brookwood Cemetery in Surrey. The service remained in operation until 1941 (Courtesy of John Clarke)

▼ The Derby and District Funeral Company Limited operated from Normanton Road in Derby. John Twells was the managing director in 1924, when the editor of the BUA Monthly described his business as, 'A Model Funeral Directing Establishment.' After training in London with John Sherry, father of Henry Sherry, John Twells returned to Derby where in 1899 he purchased a small business. On the left is the doorway into the private office, while the wide passageway gave access to the yard where the waiting room and chapels of rest were located. In quite an enterprising move for the time, there was also a coffin showroom. Some 16 black stallions were kept in the stables at the end of the yard (BUA Monthly)

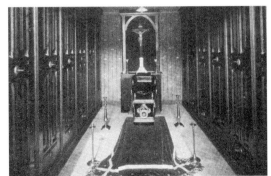

▲ This photograph of the chapel of rest at W Garstin & Sons of Wigmore Street in Marylebone dates from the 1930s. Some of the stained-glass panels have been preserved and can now be found at J H Kenyon's premises in Westbourne Grove.

◄ An early view of the chapel of rest in the premises of the south-east London firm of Francis Chappell. Here, the coffin rests on a full-size bier rather than trestles.

▼ (Left) When towards the end of 1930 Henry Smith opened chapels of rest on his premises on Battersea Park Road, the BUA Monthly was effusive in its praise:

> 'It is no new thing for undertakers to have private mortuary chapels, but it is a very commendable feature of the present era that more and more undertakers are going to the expense of providing this service.
>
> In the age of flats, tenements and apartments, the laying out of the body becomes difficult, and those in charge are seriously troubled for means to treat the departed in a decent and reverent manner. The extension of private mortuary chapels by undertakers solves the problem.
>
> The spirit in which most progressive undertakers have gone about the design of these chapels is a matter for congratulations from the great majority ... of those that we have viewed have been planned with charming restraint.
>
> Mr Henry Smith of Battersea, London, has recently had a private mortuary chapel erected on his premises ... It is a very handsome building with half-panelled and half-tiled walls, a panelled ceiling, and mosaic floor. The doors ... have leaded stained-glass panels. A feature of this chapel is the care in which Mr Smith took in designing it. The use of tiles and mosaic floor ensure absolute cleanliness ...' (BUA Monthly September 1930)

▲ (Right) The chapel in the premises of A G Hurry at St John's Wood in north London. The interior was designed by his son, Leslie Hurry, a talented theatrical designer who trained at the Royal Academy School of Painting. The Hurry family were involved in funeral service in London over many years. As earlier chapters have recounted, James Hurry was an undertaker in Stratford before becoming the long-standing secretary of the BUA. Thomas Hurry had a funeral business in the Waterloo area and received instructions concerning the coffin of Lord Kitchener (see chapter 9), whilst another branch of the family traded in St John's Wood (*BUA Monthly*)

▼ During the 1920s and 1930s, a number of British undertakers embarked on educational visits to the United States where they inspected funeral homes and attended conferences. Influenced by what they encountered, some returned home to construct purpose-built funeral homes with a fully equipped embalming theatre and chapels of rest. The Scales Funeral Service in Blackburn is one such example. Opened in 1940, this enormous building with its Art Deco frontage occupied a large site on Darwen Street. The building was demolished in the 1980s *(TUJ)*

▼ In November 1937, the Richmond-based firm of T H Sanders opened their purpose-built funeral home on Kew Road. The building to the left comprises two chapels of rest and a service chapel with staff offices above; to the right are the funeral arranging offices. Of the elegant design *The Richmond and Twickenham Times* commented,

> 'From outside, the brickwork of the new building harmonises with the older part, while the central tower and the plain cross in the brickwork are reminiscent of church architecture . . . Every night a warmly tinted light shines from the tower, being electrically regulated to start earlier as the days grow shorter, so that the dead will not seem to have been forgotten and left in darkness'.

This image dates from the end of 1937 when T H Sanders acquired a Rolls Royce fleet from J C Clark Ltd of Shepherd's Bush *(BUA Monthly)*

A Superb Fleet of Cars

FIRST DELIVERIES OF THE NEW FLEET OF ROLLS-ROYCE BEING SUPPLIED TO MESSRS. T. H. SANDERS & SONS, LTD., OF RICHMOND, SURREY

J. C. CLARK, Ltd.
GREENSIDE WORKS, GREENSIDE ROAD, GOLDHAWK ROAD
SHEPHERDS BUSH, LONDON, W.12
SHEPHERDS BUSH 5222

★
SPECIALISTS IN THE PRODUCTION
OF FUNERAL COACHWORK

▼ William Nodes opened his large funeral home at Crouch End in north London in 1937. Although only comprising offices, coffin display room and service chapel – preparation room, coffin workshop and garages were at a different location – this imposing building was designed to match the progressive architectural vocabulary employed in the immediate vicinity, particularly the Scandinavian-inspired town hall. Externally, the ground floor was clad in black granite, while the interior was panelled with walnut. A driveway at the side of the building eased the loading of the coffin onto the hearse. The building closed as a funeral home in the 1980s before becoming a public house; it is now a cycle shop.

◄ In some cases, premises could not be internally reordered to accommodate a chapel of rest. William Nodes build this small chapel in the rear garden of their premises on Bounds Green Road. The south-west London firm of Frederick W Paine also constructed chapels in the rear grounds of their branches at Sutton and Worcester Park

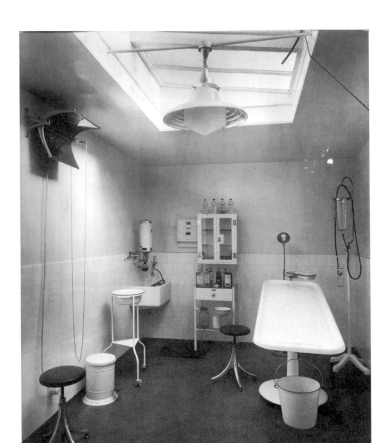

◄ A photograph taken around the late 1930s of the preparation room in William Nodes's premises at Crouch End. This facility was considered state of the art, with its walls part-tiled and part-painted, an enamel embalming table, an 'Ascot' to dispense hot water; and instruments kept in a sterilised cabinet. On the tripod stand is a valentine jar for the gravity injection of embalming fluid. The room was also used for teaching students enrolled in the Metropolitan School of Embalming, of which Mr Nodes was the principal

▼ A view taken around the 1940s of the interior of Blake and Horlock's chapel of rest at Enfield in north London. The casket is a Garrett design from Henry Smith's range

COFFINS

DOTTRIDGE BROS., LTD., UNDERTAKERS' MANUFACTURERS & WAREHOUSEMEN,

BRASS FINISHING AND ENGRAVING

DORSET WORKS, EAST ROAD, CITY ROAD, LONDON, N.

▼ ▲ Undertakers relied on wholesale suppliers for a variety of requisites such as coffins boards and sets, finished coffins and caskets, furnishings, linings and shrouds; and equipment such as trestles, palls and coffin biers. The two most important wholesale suppliers firms were Dottridge Bros and Ingall, Parsons, Clive & Co.
Founded in the 1840s, Dottridge Bros occupied a large site on the City Road at Hoxton where they constructed a wide range of wooden and metal coffins and caskets. They also made soft goods, distributed embalming fluids and equipment, built hearses and were carriagemasters to London undertakers.

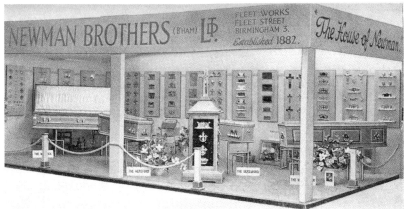

Our Stand N.A.F.D. Exhibition Blackpool 1951

▼ ▲ Many of the firms that made metal handles and furnishings were based in Birmingham, such as Newman Bros, which was founded in 1882 by Alfred Newman (FSJ). Others included Gordon and Munro, Chinn & Co, Henry Bailey and Samuel Heath. One important south London coffin supplier and carriagemaster was Henry Smith. Born in Chelsea in 1854, Smith moved to Battersea Park Road in 1869 where his business remained until 1952 when the site was compulsorily purchased. By the 1920s, his premises had an extensive plant of machinery to produce finished coffins, mouldings and coffin sets. Another London supplier of coffins and boards was J T Wybrow & Sons, whose large factory was at Loughborough Junction near Brixton (*TNFD*)

▲ Until the twentieth century, all coffins would have been constructed by hand. Oak and elm boards would have been purchased, dried and then prepared after the undertaker called at the house to measure the body. This photograph from around 1905 shows the scene in the workshop of W S Bond at Shepherd's Bush in west London. Hand saws and a plane are clearly visible

▲ A view taken in 1924 of cloth-capped employees in the workshop of the Derby and District Funeral Company. The facility was fully automated with a planing machine, circular saw and a sand-papering machine that were all electrically powered. Around 150 coffins were kept in stock and regularly used to supply country undertakers. In addition, a range of over 900 shrouds was maintained along with brass, copper, silver and oxidised coffin furniture (BUA Monthly)

◄ Around 1906, Henry Smith installed a 60hp gas engine to drive the machinery in his premises at Battersea Park Road. An under-floor belt system provided power for a device with eight small circular saws for curfing the sides of the coffins, whilst a large planing machine was used for grooving and tonguing. Another machine produced coffin mouldings. By the early 1920s, Henry Smith employed over 70 men and often kept around 3,000 coffins of all sizes and qualities in stock for immediate despatch to customers *(BUA Monthly)*

◄ This steam-driven bench saw was advertised in 1913 *(TUJ)*

No. 230.
2⅝ in. × ⅞ in.

No. 643.
2⅜ in. × ⅞ in.

No. 362½.
2⅜ in. × ⅞ in.

No. 25/a.
2 in. × 2 in.

No. 130.
2 in. × 2 in.

No. 120r.
2 in. × ⅞ in.

◄ Once the coffin had been constructed, mouldings would be secured to the base of the sides and to the lid. A wide range of designs could be obtained from suppliers, such as these elaborate examples produced by Dottridge Bros

◄ ▲ After the coffin had been constructed, it was ready to be french polished. Available in ceramic jars such as this one advertised in 1927, the polish would be painted onto the wood. The introduction of an automated spray gun enabled this process to be undertaken more quickly and evenly *(TUJ)*

Telegrams :
"Hectomar, Birmingham."

Nat. Telephone :
No. 1390 Birmingham.

WINDSOR

3462 REG.NO.

3481 REG. APPLIED FOR

3461/5 PATENT ARM REG. NO.

3482/2 PATENT ARM REG. NO.

3461 REG. NO.

3432 REG. Nº 473426

732/2812

JUBILEE B PLANISHED REGISTERED

3141 HP & BP REGISTERED

A VERY HANDSOME SET

REGISTER APPLIED FOR

A UNIVERSAL FAVOURITE

Prepared Coffin Sets and Mouldings
in Oak, Elm, Walnut, etc.

The Quality and Finish
of our Goods
is unsurpassed.

Specialities in Side Sheets, Robes,
Frillings, and all Inside Fittings.

▲ Once polished, the coffin would be furnished with fittings such as a nameplate, lid ornaments, wreath holders and handles. This elegant range was available in brass, nickel and zinc from C H Parsons in 1909 *(TUJ)*

▶ ▼ Throughout the twentieth century, designs have become more restrained. For example are those advertised in the 1930s by the Sheffield-based Marshall Bros and also those for caskets offered by Ingall, Parsons, Clive & Co dating from 1940 *(TNFD)*

▶ Cremation influenced the type of material used as brass and other cast metals could not be cremated, so manufacturers introduced furnishings made from light metal and also wood. In the late 1940s, Henry Smith offered a ring handle in English oak (with or without a fabric tassel), while Charles Hill & Co Ltd supplied non-metallic 'Ionic' handles as part of their 'Chilco' range. E J Sandall similarly advertised plain and embossed coffin mouldings and wooden cremation furniture *(FSJ)*

▲ To prevent leakage, the base of the coffin would be sealed with pitch. This bitumen-based substance continued to be used until the 1980s when it was replaced by one-piece coffin liners. Indesco was produced in the 1930s and could be applied with a brush or sprayed onto the wood *(TUJ)*

▼ Once fitted with handles and sealed, the interior was ready for lining. A pillow would be filled with sawdust or wood chippings, then calico used to pad the base and sides. Side sheets had the dual purpose of embellishing the interior whilst also covering the body. A ruff or frill would sometimes be placed around the top of the sides of the coffin. Made of white satin, they were often available with a coloured piping. Initially nailed or pinned to the sides of the coffin, the hand-held tacker gun made this task significantly easier when it was introduced in the 1940s *(FSJ)*

Electrify your methods!

Use the Skilsaw Portable Electric
Saw for all your Sawing.

RIPPING - CROSSCUTTING - KERFING

This Elm Set, including
Ends, was cut by a Skilsaw in

3 Mins. 40 Secs.

Kerfing the sides.

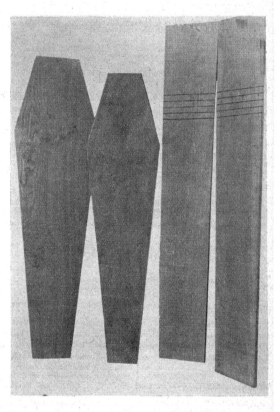

Price of Complete Machine

£23 : 0 : 0

ALL VOLTAGES. A.C. or D.C.

BRITISH EQUIPMENT CO., LTD.

RELTON HOUSE, CHEVAL PLACE, BROMPTON ROAD, LONDON, S.W.7

Phone: SLOANE 6110–6119

The Skilsaw Portable Electric Sander will do all your finishing in a
quarter of the time taken by smoothing plane, scrape and sandpaper. PRICE **£25 : 10 : 0**

▲ An increasing number of labour-saving devices specifically for the undertaker were advertised during the
1930s. The promotion of hand-held equipment such as the 'Skilsaw' along with the sander and the 'Miami'
polisher emphasised the time that could be saved; some even provided quantification *(TNFD)*

▲ The preparation of nameplates also became easier during the interwar years through the introduction of specialist equipment. In the past, they would have been prepared by a local signwriter or the details would have been handwritten onto the lid. Alternatively, the inscription could have been sketched on with a crayon before it was varnished. It would then be dusted with bronze powder; an hour later, the surplus powder would be removed to reveal the inscription. Brass nameplates would have been sent to a professional engraver. The first pantograph engraving machines were introduced in the early 1930s. Metal stencil letters were aligned on a frame before the operator would trace over the outline with a pointer attached to a revolving needle that incised the plate. A steady hand was required as one slip would result in the whole plate being ruined *(TUJ)*

▼ Throughout the twentieth century, a wide range of coffins made from a variety of materials were available. Coffins for a particular purpose were occasionally advertised. Dating from the late 1880s, the 'New Sanitary Shell' was obtainable from J Yates & Co in Manchester and continued to be marketed into the twentieth century. In the days before embalming was carried out and the body remained at home in the interval between death and the funeral, decomposition was inevitable, so this coffin's 'patent self-activating deodorising chamber' would have helped to alleviate unpleasantness. However, it was a cumbersome device and it is doubtful if many were used.

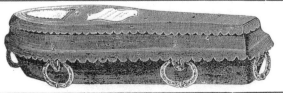

THE O.K. BUCKHOUT CHEM. CO., LTD.

One of our many Specialities.

WICKER AMBULANCE BASKET FOR REMOVAL OF BODIES.
Light, Flexible, Convenient, and Aseptic.

Large Size, with Cover, 52s. each.
With Cover and Straps, 60s. each.

We are the Head-quarters for all

Embalmers' and Disinfectors' Supplies and Undertakers' Specialities.

Manufacturers of the Famous Brands of "O.K." & "Special" Embalming Fluids.

General Agents for the Gleason Folding Laying-out Table.

Patent Automatic Coffin Lowering Devices.

Publishers of the "ART & PRACTICE OF EMBALMING."
By PROF. F. A. SULLIVAN, Price 20s., post paid.

SEND FOR ILLUSTRATED CATALOGUE AND PRICE LIST.

Note New Address :—
43, Grafton Street, Tottenham Court Road, London, W.

▲ Advertised in 1903, this wicker 'ambulance basket' was intended for removals, rather than more contemporary 'green' burials *(TUJ)*

▶ Wooden caskets have long been available from suppliers. This is an elegant example produced in 1908 by Dottridge Bros *(TUJ)*

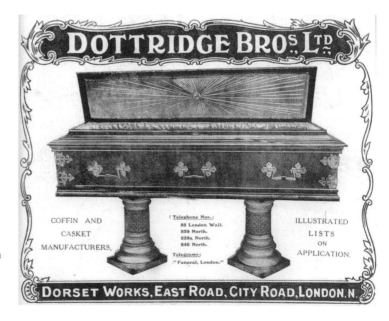

▲ One labour-saving way of coffin making was by using a pre-prepared set comprising the sides, lid, base, head and foot end. Such coffins could be promptly despatched from Dottridge Bros or Ingall, Parsons, Clive & Co following receipt of a measurement from the undertaker. On arrival at the destination they would be assembled, sealed, lined and fitted with handles and nameplate *(TUJ)*

Telegraphic Address : "IMMEDIATE, BIRMINGHAM."   Telephone : 1968 MIDLAND.

PATENT METALLIC AIR-TIGHT SHELLS.

THE PATENT METALLIC AIR-TIGHT COFFIN CO., IN SUBMITTING THE ABOVE TO YOUR NOTICE
BEG TO POINT OUT THE FOLLOWING ADVANTAGES OVER ALL OTHER SHELLS.

They are very strong, light, and a good shape ; are guaranteed perfectly sound and air-tight, and are the only Shells that can be thoroughly tested.
They are not wood shells covered with zinc, but are all strong metal, non-corrosive, and are practically imperishable.
These Shells were introduced about thirty years ago, and several Patented improvements have since been added. Their sale is constantly increasing
as where once used they are always recommended.
A very large stock of all sizes is kept, and a Shell or Coffin can be sent off within an hour of receipt of telegram or letter.
The Shells are sent out fitted with best flannel linings and pillows complete, and, if required, a plate of glass can be inserted in the lid without in the
least destroying its efficiency.

MEDICAL PRACTITIONERS, having carefully inspected the Patent Metallic Air-Tight Coffins and Shells manufactured by this Company, recommend them very
strongly upon sanitary grounds. The formation and diffusion of the noxious gases resulting from decomposition are entirely prevented by the non-access of
atmospheric air, and this even in the hottest weather. This is at all times an important desideratum, and becomes, in the case of certain infectious disorders, a
most valuable means of preventing their further spread.
In cases of death from Dropsy, &c., the fact of Wooden Coffins being porous often leads to results which are, at all events, highly disagreeable. These considerations, in conjunction with the elegance, lightness, and economy of the Metallic Coffins, must, in no great time, ensure their universal adoption.

THE PATENT METALLIC AIR-TIGHT COFFIN CO.,
70, Broad Street, BIRMINGHAM.

◄ Metal-lined coffins would have been necessary for interments in catacombs, vaults and brick-lined graves, and also for overseas transportation (*TUJ*)

◄ A metal casket advertised in 1932 (*TUJ*)

SOUDAN

◄ The 'Soudan' was one of a large range of coffins and caskets supplied by Henry Smith. Many were named after streets in the Wandsworth and Battersea area

Code: "ABBEY."

COFFINS
AND
CASKETS

THE supply of Funeral Furniture is always a matter of urgency. We can supply you with Coffins of the Finest Timber, Workmanship, and Finish, in the White or Hand French Polished.

Despatched the day the order is received.
PLEASE SEND FOR ILLUSTRATED CATALOGUE.

FOX & CHARLES, *Wholesale Manufacturers* **31-33, Bristol St., BIRMINGHAM, 5**
Telegrams: "FOXCHARCO, BIRMINGHAM."   Phone: MIDLAND 2068.

▲ The 'Abbey' coffin with its raised lid and cross as supplied by Fox and Charles in 1933 *(TUJ)*

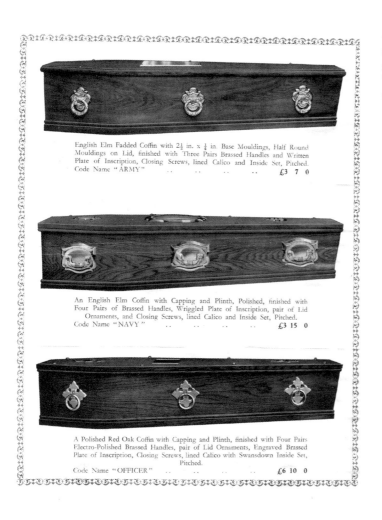

English Elm Fadded Coffin with 2½ in. x ¾ in. Base Mouldings, Half Round Mouldings on Lid, finished with Three Pairs Brassed Handles and Written Plate of Inscription, Closing Screws, lined Calico and Inside Set, Pitched. Code Name "ARMY" £3 7 0

An English Elm Coffin with Capping and Plinth, Polished, finished with Four Pairs of Brassed Handles, Wriggled Plate of Inscription, pair of Lid Ornaments, and Closing Screws, lined Calico and Inside Set, Pitched. Code Name "NAVY" £3 15 0

A Polished Red Oak Coffin with Capping and Plinth, finished with Four Pairs Electro-Polished Brassed Handles, pair of Lid Ornaments, Engraved Brassed Plate of Inscription, Closing Screws, lined Calico with Swansdown Inside Set, Pitched. Code Name "OFFICER" £6 10 0

◀ A page from a Dottridge Bros coffin brochure. Considering the names of the coffins – 'Army', 'Navy' and 'Officer' – it probably dates from the 1940s

★
Something new in Coffins

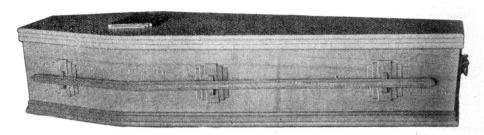

THE BERESFORD OAK VENEERED COFFIN

A COFFIN OF OUTSTANDING QUALITY

These coffins are constructed of elm in the usual way then veneered with selected highly figured oak. They are both light and strong and can safely be put into stock. All moulds are oak. Each coffin is made by experts in this class of work and cannot be distinguished from solid oak. Sold with or without furniture in the white or waxed.

Extract from a letter. *The Beresford coffin has indeed come above all expectations. It was not a coffin as we know them but a perfect cabinet. No more Elm for us so long as you can supply them.*

Write for price and particulars and your nearest agent to The Sole Manufacturers :

J. NICHOLSON & SONS, Longlands Works, Windermere

Telephone : WINDERMERE 475

▲ Veneered coffins started to appear in the late 1930s. The 'Beresford' coffin dates from 1941 and was ideal for cremation as all the furnishings were made of wood, including the long bar handles *(TNFD)*

CHELSEA

MORTLAKE

BROMPTON

◄ To meet the increasing demand for cremation in the 1940s, many suppliers offered a wide range of wooden caskets for ashes. These are a few obtainable from Henry Smith

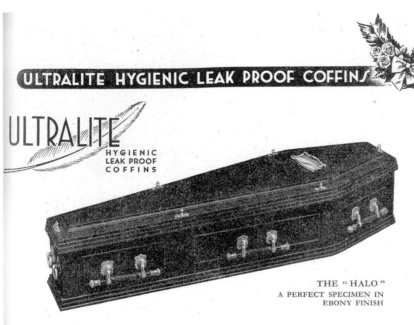

ULTRALITE HYGIENIC LEAK PROOF COFFINS

ULTRALITE
HYGIENIC
LEAK PROOF
COFFINS

THE " HALO "
A PERFECT SPECIMEN IN
EBONY FINISH

MANY years of scientific research have produced a masterpiece of coffin construction—bakelite moulded under immense pressure, giving tremendous strength with remarkable lightness in weight—the average being only 92 lbs.

There are only two pieces—the box and the lid, each fused in one complete perfect moulding ; there are therefore no joints, no possible warpings or leakages under any conditions, and all imperfections of wooden coffins are for all time removed. Each of the two separate pieces will easily stand a strain of five cwts., and withstand far more rough usage than the usual handling.

ULTRALITE HYGIENIC MOULDED COFFINS are beautiful in design and have a lustrous glass-like finish incorporated in the moulding which cannot be surpassed. They are made in many grains, oak, ebony, walnut, mahogany, etc., which, with the superb finish, cannot be equalled by the most expensive timbers. Their aesthetic dignity appeals to all, and their immense strength coupled with their easiness of handling commends them to the modern Funeral Director. They are also eminently suitable for cremations.

Timber is scarce—it may soon be prohibitive ; it behoves you to examine frankly and commercially the **ULTRALITE COFFIN**.

FULL DETAILS SUBMITTED ON APPLICATION.
ADDRESS YOUR ENQUIRIES TO

ULTRALITE CASKET COMPANY LTD.

EAGLE WORKS, TAME VALLEY
STALYBRIDGE, CHESHIRE

Telephone: STALYBRIDGE 2463

IMPERVIOUS TO WATER OR ACID

▲ Bakelite was developed in the early years of the twentieth century (by Dr Leo Baekeland) and in 1941, this coffin was advertised as '. . . eminently suitable for cremation'. However, its lifespan was only short as the cremation authorities at Hull complained that they gave off a '...tremendous amount of smoke.' *(TNFD)*

TRANSPORT

At the start of the twentieth century, three modes of funerary transport were regularly used: wheel biers, horse-drawn hearses and the railway network. Although all are still in occasional use, it is the motor hearse that really became the replacement. First introduced around 1900, during the interwar years undertakers gradually shifted to this new form of transport; by the 1950s, the horse-drawn hearse had disappeared

THE WHEEL BIER

The decision to use different forms of transport is largely dependent on distance to move a coffin from where it is resting, to the place of burial or cremation. For funerals taking place in rural areas and also in some cemeteries, a wheel bier would be used.

▼ This rural scene shows two men pulling a cart bier. The date is unknown, but from looking at the clothing, it is probably around the time of WWI

▼ The coffin appears to be at quite a height above the wheel bier in this image of a procession at the side of a church

◄ The coffin containing the Bishop of Chichester, the Rt Revd Ernest Roland Wilberforce, being wheeled through the cathedral grounds on 14 September 1907. The son of a bishop, he was born in 1840 and educated at Oxford. Before being translated to Chichester in 1895, he was the first bishop of the newly created Diocese of Newcastle

► 'Incense in Kingsway' was the headline in the *Daily Sketch* describing Father Arthur Stanton's funeral procession. A faithful curate who ministered to the poor for over 50 years, his coffin, headed by a thurifer and churchwardens, was wheeled by fellow clergy from St Alban the Martyr, Brook Street in Holborn, to the London Necropolis station at Waterloo. At 2.30 p.m. on 1 April 1913 a special train carried the coffin to Brookwood for interment in the church's own section of the cemetery. The procession down Kingsway and over Waterloo Bridge would have been quite a sight

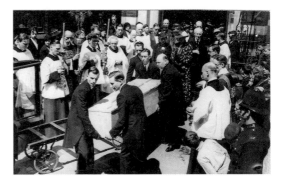

► Guided by the undertaker, with assistance from members of the Dorset Regiment, this photograph of a wheel bier was taken around the time of WW1

▼ Supplied by Dottridge Bros, these biers were advertised on the cover of *TUJ* in May 1906. The two upper designs with a compartment for the coffin indicate the origin of the word 'hearse' – as a triangular-shaped frame used as a candleholder beneath which would be placed the coffin during a service in church. The candles would be secured by prickets which resembled the teeth of a harrow. The French word for harrow is 'hearse' *(TUJ)*

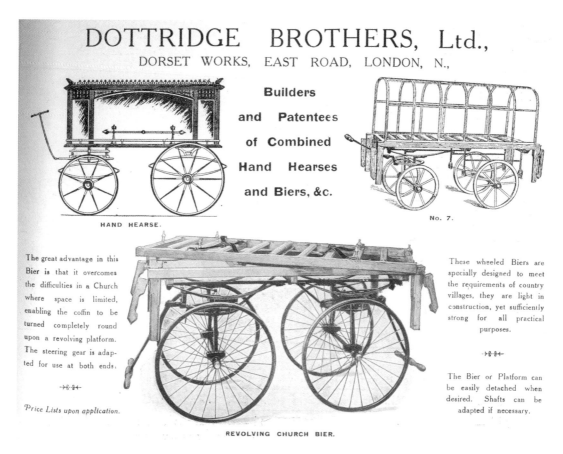

DOTTRIDGE BROTHERS, Ltd.,

DORSET WORKS, EAST ROAD, LONDON, N.,

Builders and Patentees of Combined Hand Hearses and Biers, &c.

HAND HEARSE.

No. 7.

The great advantage in this Bier is that it overcomes the difficulties in a Church where space is limited, enabling the coffin to be turned completely round upon a revolving platform. The steering gear is adapted for use at both ends.

Price Lists upon application.

These wheeled Biers are specially designed to meet the requirements of country villages, they are light in construction, yet sufficiently strong for all practical purposes.

The Bier or Platform can be easily detached when desired. Shafts can be adapted if necessary.

REVOLVING CHURCH BIER.

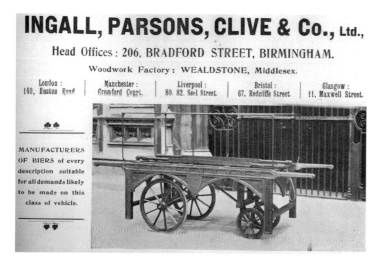

INGALL, PARSONS, CLIVE & Co., Ltd.,

Head Offices : 206, BRADFORD STREET, BIRMINGHAM.

Woodwork Factory : WEALDSTONE, Middlesex.

| London : 140, Euston Road | Manchester : Cromford Court, | Liverpool : 80. 82. Seel Street. | Bristol : 67, Redcliffe Street. | Glasgow : 11, Maxwell Street. |

MANUFACTURERS OF BIERS of every description suitable for all demands likely to be made on this class of vehicle.

▲ Ingall, Parsons, Clive & Co advertised this substantial gothic-style wheel bier in 1928 *(TUJ)*

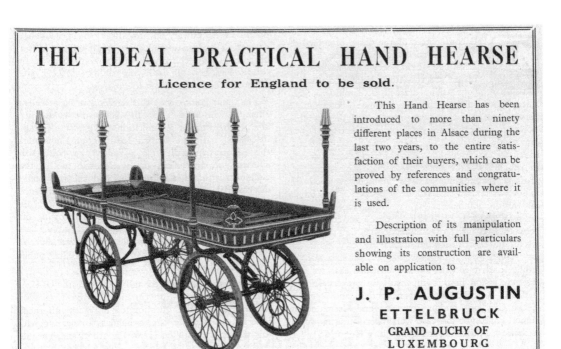

THE IDEAL PRACTICAL HAND HEARSE

Licence for England to be sold.

This Hand Hearse has been introduced to more than ninety different places in Alsace during the last two years, to the entire satisfaction of their buyers, which can be proved by references and congratulations of the communities where it is used.

Description of its manipulation and illustration with full particulars showing its construction are available on application to

J. P. AUGUSTIN
ETTELBRUCK
GRAND DUCHY OF
LUXEMBOURG

▲ Dating from 1931, the Luxenborg was probably the last wheel bier to be advertised to undertakers (*BUA Monthly*)

THE HORSE-DRAWN HEARSE

▲ Over the years, many different designs of horse-drawn hearse have appeared, with the two principal firms responsible for their construction being Dottridge Bros and John Marston in Birmingham. This image shows the interior of the coachbuilding works at Dottridge Bros (*TUJ*)

JOHN MARSTON'S CARRIAGE WORKS, Ltd.,
21 to 27, Bradford St., BIRMINGHAM.

Telegrams: "HANSOMS, BIRMINGHAM."   Telephone: No. MIDLAND 1776.

Designers, Patentees, and Builders of Hearses, Funeral Cars, Clarence Coaches, Undertakers' Broughams, Handy's Patent Hearses, and every description of Carriage used in the Trade, Also Motor Hearses, Funeral Buses, and every kind of Motor Car and Van Bodies.

REVISED PRICES

BEST DISCOUNT FOR CASH.

TERMS ARRANGED TO SUIT CUSTOMERS.

IMPROVED GLASS CAR.   Various Sizes to suit, Single, Pair, or Four Horses.

SPECIAL LIGHT GLASS HEARSES from £85.
Write for Drawings, Photographs, and Full Particulars, Free.

LATEST PATENT REVOLVING-FRONT SHELIBERE.

SECOND-HAND HEARSES, from £20; COACHES, £25; SHELIBERES, £30.

Our increased trade is the result of giving Customers Smart Designs, Best Material and Workmanship, Special Attention and Supervision in all details and finish, which has merited their recommendation to others.

GOOD STOCK MEANS INCREASED TRADE AND REDUCED EXPENSES

▲ Two Marston hearses advertised in 1913. The lower hearse has a revolving coffin compartment. Other coachbuilders included Lewis, Hammond & Co who supplied 'Funeral carriages and hearses of every description' from their showroom on Harrow Road in Paddington. Founded by Thomas Hammond in 1879, he came from a family in the north of England long involved with carriagebuilding. In 1903 they advertised their 'Eclipse car [hearse].' E T Wainwright of Eagle Carriage Works, Aston Road in Birmingham also constructed the latest designs of the Washington cars and hearses *(TIJ)*

To Funeral Carriage Proprietors.

GREAT
AUTUMN
SALE
By Auction

OF
Superior
BLACK
HORSES

80 BLACK STALLIONS & GELDINGS

Elephant & Castle Horse Repository,

NEW KENT ROAD, LONDON, S.E.,

Will hold their next SALE for Undertakers'
Property at the Repository on

WEDNESDAY, OCTOBER 1st, 1913,

Commencing at 1 o'clock.

There will be included

80 Black Stallions and Geldings,

Glass Side Hearses, Mourning Coaches, Broughams,
and about 40 Sets of Single & Double Harness.

— ALSO —

30 Fresh Imported Black Stallions
and Geldings,

The Property of Mr. W. E. Chapman, "Prince of Wales" Yard,
Union Road, Clapham, S.W.

These Horses are of Superior Quality, all from
three to five years old, sound in wind and eyes.

On View Day Prior.   Catalogues forwarded on application.

GEORGE STEWART, Manager.

◄ One of the main suppliers of horses to undertakers was The Elephant & Castle Horse Repository located on the New Kent Road in south London. They also auctioned horses and horse-drawn vehicles. This advertisement for a sale of horses dates from 1913. W E Chapman of Wandsworth imported black stallions and geldings, as did J Schrall of Pentonville. In her study of horse-drawn transport in London, historian Sally Child describes those used by undertakers as being part of 'The Black Brigade' and notes a tradition found at Dottridge Bros' stables, as recounted by W J Gordon in *The Horse World of London* (1893):

> *'Over every horse was his name, a curious assortment of celebrities, ancient or modern. An ideal team at the time consisted of Bradlaugh, John Knox, Dr Adler and Cardinal Manning. But the practice of naming horses after church and chapel dignitaries is being dropped owing to a superstition of the stable. All the horses named after that sort of person seemed to go wrong somehow.'*

▼ The use of horses by undertakers gave rise to an allied industry of suppliers. J W & T Connolly of King's Cross offering patented rubber tyres, which were described as the 'best on the market', while harness oil was offered by the Frank Miller Company of Moorgate. S & H Harris similarly supplied saddle paste, saddle soap, plate polishing paste and black dye. Uveco's advertisement is from 1907 *(TUJ)*

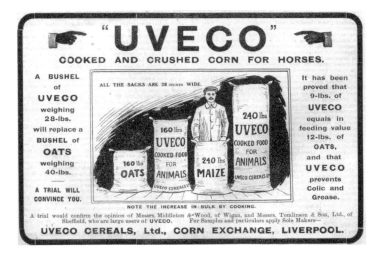

▶ A closed hearse used for removals owned by B Higginbottom of Farnworth near Bolton in Lancashire. He offered, 'Express removal of bodies to and from all parts of the country by motor, rail or van.'

▶ This closed-sided hearse poses outside Frederick W Paine in Kingston and dates from 1907

▶ A Marston hearse advertised in 1913. Design No 11161 was particularly decorative, including a central crown and anchor in the centre of the roof (a motif found on many hearses) and a seat supported by intricate ironwork *(TUJ)*

MARSTON'S NEW DESIGN CAR. No. 11161.
This pattern of Hearse is built in various sizes to suit different districts.

▲ A 'floral carriage' built by Dottridge Bros in 1909. The position of the carriage is often a subject of debate: should it take precedence over the hearse? Dispensing with the etched glass and creating an open-sided hearse was considered to be the simplest design and matched the reforming agenda of the late Victorian period. William Garstin, the central London undertaker and pioneer of funeral reform, had one constructed in Paris and started using it in 1855.

▲ A floral hearse outside the church of St Leonard Bromley by Bow, London

▲ An open-sided hearse used in October 1910 at the funeral of Thomas Loates in Brighton

▶ An important change came in January 1914 when the Royal Society for the Prevention of Cruelty to Animals forbade the use of black ostrich plumes: when wet, they were particularly heavy and put a strain on horses' necks. In the January edition of the RSPCA's periodical *The Animal World*, the success of their appeal was indicated:

> 'With the coming of the year 1914 there passes away from general use that old-fashioned appurtenance of mourning procession, the funeral plume. Like the mutes with the bands, it has served its day. A relic of the imposing procession of antique times, its presence was an anachronism, and its disappearance will occasion little regret. From the point of view of humanity to animals indeed its disappearance is to be applauded...'

The article then traced the steps taken to reach this position, including representations to the British Undertakers' Association. They noted unanimous support from the trade and were profuse in their appreciation:

> 'This signal success is, of course, greatly to the credit of the RSPCA, but the wonderful unanimity and assistance of the undertakers themselves must not be forgotten. In their eminently sane attitude, they have proved themselves to be a most reasonable and humane body of traders, and the thanks of the Society and of all humane persons are gratefully accorded them for their sympathy and co-operation.'

During the interwar years, the horse-drawn hearse still had a presence on the streets, but it was clear that they were soon to be replaced by motor power. In the period up to the late 1930s, many undertakers offered both modes of transport to clients, but it was clear that the motor hearse was increasingly preferred. However, the horse-drawn hearse's absence from the streets was not for long as, in the 1980s, they made a reappearance

ROYAL SOCIETY

FOR THE

PREVENTION OF CRUELTY TO ANIMALS.

Notice to Undertakers.

FUNERAL PLUMES ON HORSES.

Investigations made by the above Society have established beyond doubt the fact that the use of Funeral Plumes on horses often entails suffering and discomfort to the animals wearing them, and at a meeting of the British Undertakers' Association (London Centre), held on the 19th May, 1913, at the Farringdon Restaurant, Farringdon Street, E.C., the plume was unanimously condemned by the members then and there present as cruel and unnecessary.

> Section 1 of The Protection of Animals' Act, 1911, forbids the infliction of "any unnecessary suffering" upon any animal, and provides that persons offending in this respect may be fined or imprisoned.

Under the above circumstances the Society strongly recommends that the practice be discontinued.

E. G. FAIRHOLME, Chief Secretary.

105, Jermyn Street, London, S.W.

Mr. W. G. Spicer, Sec. B.U.A., asks us to give publicity to the above Notice which explains itself.

Copies of this Notice will be sent to every undertaker in London and adjoining districts by the Society.

Owing to the receipt of the above notice,

We have agreed to DISCONTINUE THE USE of the FUNERAL PLUME AT THIS ESTABLISHMENT on and after the 1st January, 1914.

▲ Funerals of a multiple of victims posed no problems. Depending on the number of coffins, different modes of transport can be employed, such as the lorry for victims of the R101 and R38 airship disasters. In this image, two coffins are placed side by side in the compartment of the horse-drawn hearse.

▼ The undertaker and hearse driver wear white gloves and white weepers around their hats at this funeral of a child. The horse palls are similarly decorated.

▼ A horse drawn stands outside the long-established Birmingham firm of Harman and Bastock.

THE MOTOR HEARSE

▾ Much debate concerns which undertaker was first to use a motor hearse. Although Reuben Thompson of Sheffield has this distinction, his Wolseley vehicle was used in 1900 for the distant deliveries of coffins rather than for funerals *(TUJ)*

▸ Henry Smith's hearse was similarly used for conveying a coffin over a long distance, in this case from London to Eastbourne in May 1904: a charge of £10 was made. The coffin was completely enclosed in a box upon which the bearers sat *(TUJ)*

▴ In January 1910 *TUJ* commented that, 'The motor is entering so much into our daily life that it was inevitable that it should be eventually called in to serve the undertaker in his dealings with the dead.' The article goes on to reveal that the first documented use of a motor hearse for a funeral was by Councillor Pargetter of Coventry. The hearse was constructed in the firm's own workshop using a Lotis chassis with a 16/20hp engine. Lotis Sturmey Motors was based in the city and made taxicabs. The body of the hearse was a Washington-Car type with glass sides and curtains. It was first used in December 1909 when three distance removals were carried out in one week. One of the strongest reasons put forward for its adoption was '...to do away with the great inconvenience connected with railway transit, as well as the objection of having the bodies conveyed in milk, fruit and other vans, used by railway companies for this purpose.' By July 1911, the hearse had completed over 10,000 miles *(TUJ)*

Many hearses, such as that owned by C E Hitchcock of Plaistow, were an enclosed hearse for removals rather than for funerals. These were referred to as a 'handy' – a term first used around 1907

The Pioneers of Modern Hearse Building.

Get a MARSTON built HEARSE or CAR and you have the BEST and most UP-TO-DATE for HORSE or MOTOR traction. The Best is the Cheapest.

Combined Hearse and Passenger-Carrying Saloon on Humber Chassis.

No. 12074. Glass-sided Hearse with enclosed front on Vulcan Chassis.

The latest, neatest, and most approved designs in Motor Hearses complete, or Bodies only fitted to Customer's own new or second - hand chassis, any make. : :

Best Discount for Cash or Special Terms arranged : to suit Customers. :

See Testimonials from Customers : using our Hearses.

Improved Glass Hearse on Motobloc Chassis.

No. 11972. High-Class Combined Saloon Hearse and Passenger Carrying Motor on Daimler Chassis.

Combined Saloon Hearse and Ambulance on Napier Chassis.

JOHN MARSTON'S CARRIAGE WORKS, Ltd.,

Telegrams: "HANSOMS, BIRMINGHAM."

21 to 28, Bradford Street, BIRMINGHAM.

Telephone: No. MIDLAND 1776.

SPECIAL LIGHT GLASS HEARSES from £85.

SECOND-HAND HEARSES, from £20;   COACHES, £25;   SHELIBERES, £30.

▲ The carriagebuilders soon responded to the shift towards motor vehicles, which started in earnest around 1910–1914. From this latter date, John Marston described themselves as, 'The Pioneers of Modern Hearse Building – Designers, Patentees, and Builders of Hearses, Funeral Cars, Clarence Coaches, Undertakers' Broughams, Handy's Patent Hearses, and every description of Carriage used in the Trade. Also Motor Hearses, Funeral Buses, and every kind of Motor Car and Van Bodies.'

The advent of motor vehicles ushered in an opportunity to make designs less elaborate. Although the first motor hearses were literally the coffin compartment of a horse-drawn vehicle secured onto a chassis, they soon became purpose-built and streamlined *(TUJ)*

▲ The Brighton firm of Attree and Kent Ltd introduced their motor hearse built by Messrs T Hammond of Paddington in 1913 *(TUJ)*

▶ J H Kenyon advertised this motor hearse for hire work in 1914. The desire for the ostentatious hearse, both horse-drawn and motorised, was gradually diminishing around the time of WWI. By the end of 1918, carriagemasters in the urban areas quickly recognised the potential of possessing a modern motorised fleet

▶ These two images show the change of use of Henry Smith's premises from stable (1906) to garage (1925) *(TUJ* and *BUA* Monthly)

▲ This image shows Frederick Paine standing in front of his office at 24 London Road (now Old London Road), Kingston, with what is believed to be the first motor funeral in the district. The date is 1913. An increasing number of coachbuilders were supplying the market for motors hearses as these advertisements indicate.

▲ A Ford hearse built by Dottridge Bros in 1922 *(TUJ)*

▲ The Wigan firm of Middleton & Wood advertised their hearse in the BUA Monthly in 1924 *(BUA Monthly)*

MAXWELL BROS., Ltd., 312 Brixton Road, 252 Brixton Hill, and 143 High Road, Streatham.

MAXWELL BROS., Ltd. 312 Brixton Road 252 Brixton Hill, and 143 High Road Streatham.

▲ The lower picture of Maxwell Bros' hearse shows the ingenious wheel bier arrangement, including the portable tracks for ease of movement

The Hearse illustrated is a unique combination of cabinet work and coach-building. The interior floor is of waxed oak parquetry. The roof is of carved oak and the finishing pieces are veneered instead of being painted.

Illustration shows G.M.B. Motor Hearse with glass embossed sides.

Arrangements will be made for any Funeral Furnisher to inspect this Hearse upon request. Don't hesitate to write; there is no obligation. Art Brochure will be sent post free.

FIELDING & BOTTOMLEY Union St. South, Halifax, Yorks. Phone 2634.

There's Fifty Years Reputation behind this Fielding & Bottomley Hearse!

Reputations are too valuable to be lightly thrown away for a little extra profit, and that is why Fielding & Bottomley pay such strict attention to each detail in the construction of their Motor Hearses.

The result is that they can guarantee satisfaction. The design is also dignified and calculated to give a decided "pull" to Undertakers who possess these vehicles.

◄ The coffin compartment of Fielding and Bottomley's hearse advertised in 1929 was very similar to that drawn by a horse *(BUA Monthly)*

ARMSTRONG SIDDELEY
30 H.P. SIX CYLINDER CARS.
Finest Value in Fine Cars.

THE 30 h.p. Armstrong Siddeley 6-Cylinder is a distinguished-looking motor carriage of perfect comfort and ample power at an exceptionally moderate price. There is always a reserve of power, and it carries the maximum number of people (the car illustrated will seat 7) in complete comfort and silence. Thousands of owners vouch for its efficiency.

Supplied as Chassis only—or with body to suit special requirements.

PRICES : Chassis, £700. Open Touring Car (De Luxe), £950. Touring Landaulette, £1,050. Limousine or Three-quarter Landaulette, £1,125. Enclosed Limousine or Landaulette, £1,350. Front Wheel Brakes, £35.

Write for Booklet "H.3" and address of nearest Agent.
ARMSTRONG SIDDELEY MOTORS LIMITED, COVENTRY.
(Allied with Sir W. G. Armstrong Whitworth & Co., Ltd.)
Manchester : 33, King Street, West. London : 10, Old Bond Street, W.1. Service Depots at Coventry, London, Manchester, Newcastle, Glasgow, Leeds, and Bristol.

The 30 H.P. Armstrong Siddeley 6-Cylinder Enclosed Landaulette. Price £1,250 complete.

YOU CANNOT BUY A BETTER CAR.

◄ Costing £1,250 in 1925, this Armstrong Siddeley limousine represented quite an investment for the undertaker *(BUA Monthly)*

NEW HEARSE BODIES from £130 to £350.   **The greatest value ever offered.**

Any type of hearse body built to order at competitive prices. Designs, photographs and specifications on application. Renovations and repairs to Horse-drawn and Motor Funeral Coaches and Hearses, Painting, Lining, Silvering, etc. and General overhaul at prices much lower than are usually charged for this class of work. All kinds of conversions including Limousines to Hearses.

Dual Purpose Body

Special Design Four-door Saloon Hearse. Various Others

ARMSTRONG & Co. (Coachbuilders) **Ltd.,** 4 Leysfield Road, Shepherd's Bush, London, W.12.
'Phone: Shepherds Bush 1577.

◄ An Armstrong & Co advertisement dating from October 1930. By the 1930s, hearse bodies had become streamlined, ushering in an age of sleek elegance *(BUA Monthly)*

ROLLS ROYCE HEARSE 87" - BIER behind driver Chromium fittings Cellulose finish

BODY BUILT BY BLAGG ON 40/45 ROLLS ROYCE CHASSIS

PERFECTION realised in HEARSE BODIES by BLAGG

◄ A cortège comprising Rolls Royce vehicles certainly made heads turn. This hearse was built in 1934 by Blagg & Co of Sunderland *(BUA Monthly)*

▲ Alpe & Saunders Ltd had a showroom on Brompton Road, while their coachbuilding works were at Kew Gardens. Hearse designs using Armstrong Siddeley, Austin six, Rolls Royce and Daimler are shown in this 1938 advertisement

▲ A 1930s Daimler hearse, possibly constructed by Alpe & Saunders, parked outside the rear entrance of an undertaker's

▲ A 1939 advertisement for Lancefield Coachwork. Within a few years, they would stop constructing hearses as their engineering equipment and expertise was requisitioned for the war effort (BUA Monthly)

◄ As it was not appropriate to convey a child's coffin in a motor hearse due to the small size of the coffin (and securing it to the deck), the issue was solved by creating a compartment between the driver and passengers in a limousine. This vehicle was made for W English & Sons of Bethnal Green and dates from 1940 (BUA Monthly)

▲ Hearses under construction on the production line at Woodall's Halifax premises, just before the outbreak of WWII *(BUA Monthly)*

▲ A 1950 Austin Sheerline hearse available from Arthur Mulliner coachbuilders of Nottingham *(BUA Monthly)*

OTHER FORMS OF TRANSPORT

In addition to the wheel bier, horse-drawn and motor hearse, other forms of transportation have been used to convey coffins. Elsewhere in this book are examples of alternatives, such as a bus (page 298), tram (page 289), autocar (page 297) and farm wagon (page 305). In addition, there are images of a coffin on a gun carriage drawn by hand, such as those for the funeral of Colonel Cody (page 295). Used at royal funerals, this mode of transport has its origin in an unfortunate incident. The tradition started in 1901 when Queen Victoria's coffin had to be transported from Windsor station up the hill to the castle. It was a heavy load and the team of horses, having had to wait for hours in the bitter cold, when instructed to move defiantly stood their ground. Then one of the horses reared up, kicked, and in doing so tangled the traces attached to the gun carriage. There was no way the horses could be used. Prince Louis of Battenberg suggested that the naval guard of honour pull the gun carriage; the king approved and so a new tradition was born – one which has been used at royal funerals ever since.

▶ This image shows Bluejackets pulling Queen Victoria's coffin up the High Street in Windsor; it is seen here passing the London and County Bank

▼ Sailors pulling the gun carriage bearing Edward VII's coffin in May 1910, seen here leaving Windsor station, and then in the grounds of Windsor Castle

The London Necropolis
Company.

ESTABLISHED 1850.

NEW OFFICES *3*

AND PRIVATE RAILWAY STATION:

121, Westminster Bridge Road.

CEMETERY:

BROOKWOOD, NR. WOKING.

The largest and most beautiful
in Great Britain.

Trade Discount on Fees, 5 per cent.

Registered Telegraphic Address : "Tenebratio, London."
Telephone No. 839 Hop.

▲ The development of the railways in the nineteenth century enabled coffins to be transported efficiently both in terms of cost and speed. Undertakers would have been responsible for delivering and collecting coffins from stations. In addition to the main line service, special trains and branch lines exclusively for funerals would be established. The Brookwood Necropolis Railway which ran from Waterloo into Brookwood Cemetery is comparatively well known. However, a further example was the service from King's Cross to the Great Northern Cemetery at New Southgate, which ran from 1861 to 1863

▶ Bearers on the platform at the London Necropolis station waiting to load a coffin into a hearse van

▶ As shown in the images dating from 1915 for the funeral of a bus driver and conductor (see page 298), those killed in the course of duty often lead to colleagues participating in the funeral ceremony. This is particularly the case for the armed forces, in addition to policeman and firemen. In respect of the latter, the coffin is usually conveyed on the back of a fire engine or horse-drawn vehicle, as seen here during a funeral in Aldershot

▶ These two images show a horse-drawn fire appliance with a coffin covered by a Union flag. Both locations are unidentified

▶ A motor tender is used to convey a fireman's coffin in Ryde, Isle of Wight

▸ An Imperial Airways flight arriving at Woodford Aerodrome in Cheshire, 1927. Although road, rail and sea continued to be the most convenient mode of transport, by the 1920s aircraft were being used to deliver coffins. The first occasion when an aircraft is reputed to have been used for this task was in 1924. Walter Carlisle died in a car accident in Germany and his body was flown from Cologne to Croydon in Surrey. A hearse then took the coffin from the airport to Euston station for the train journey to the north of England, where the funeral was held. This image from August 1927 shows an Imperial Airways Vickers Vulcan aircraft that departed from Basle in Switzerland and landed at Woodford Aerodrome in Cheshire. The flight – in what was referred to as an 'aerial hearse' – took 10½ hours and the only stop was in Paris for fuel and, according to the account, for the pilot to have his breakfast. At the aerodrome, the plane was met by a motor hearse, and mechanics acted as bearers to remove the coffin *(TUJ)*

▾ Frederick W Paine's own aircraft in 1948

SHOULDERING THE COFFIN

Irrespective of the mode of transport, the bearing of coffins has been a hallmark of British funerals. The number of bearers used to carry the coffin at shoulder height has usually been six or four. This is necessary not only on account of the weight of the coffin, due, for example, to solid-wood and triple-lined coffins, but also when a distance is to be walked to the church. Some bearers link arms, or those at the head-end of the coffin link across the shoulders.

As was noted in chapter 2, an agreement between the BUA and the BFWA negotiated the use of a minimum number of bearers, and also that bearers may not also drive vehicles. The latter was especially important when managing horse-drawn vehicles; many cemeteries forbade these to be left unattended, as this sign from the City of London Cemetery indicates.

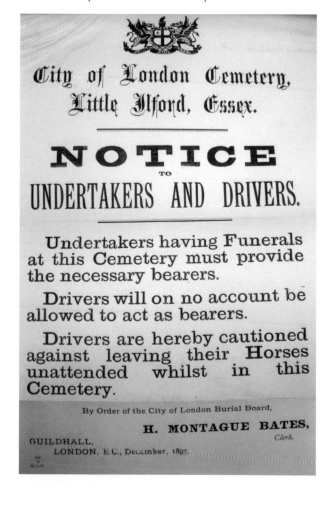

City of London Cemetery, Little Ilford, Essex.

NOTICE
TO
UNDERTAKERS AND DRIVERS.

Undertakers having Funerals at this Cemetery must provide the necessary bearers.

Drivers will on no account be allowed to act as bearers.

Drivers are hereby cautioned against leaving their Horses unattended whilst in this Cemetery.

By Order of the City of London Burial Board,

H. MONTAGUE BATES,
Clerk.

GUILDHALL,
LONDON, E.C., December, 1897.

▲ In some cases, a coffin was shouldered on a frame, which would then rest on trestles or a wheel bier as shown in this image

▶ The frame supporting Lord Northcliffe's coffin is lifted from a bier and placed on the shoulders of four bearers. This funeral took place at Westminster Abbey, and the large coffin is covered by the Abbey's pall presented by the Actors' Church Union in 1920 and used at significant funerals such as Queen Alexandra (1925) and for the lying in states of King George V (1936) and Queen Mary (1953). It can also be seen on page 133 supporting the coffin of the Unknown Warrior. When shouldering some distance or over uneven ground, the coffin was often secured to the frame with leather straps

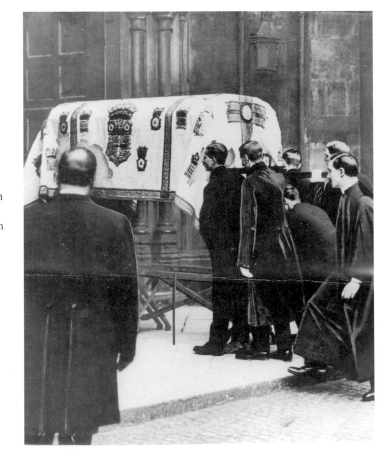

◄ Six bearers, with one each at the head and foot end of the coffin, support this well-secured coffin as it is carried across the churchyard at Ospringe in Kent

▲ A frame supported by four bearers approaches the church of St Peter and St Paul in Old Felixstowe, Suffolk at the funeral of a freemason

◄ A coffin with massive brass handles and bedecked with flowers is secured with hessian webbing to the large frame that doubles as a trestle

◄ Six bearers place this coffin onto a farm cart, elaborately decorated with foliage

▲ The ideal arrangement for bearing is depicted in this WWI image, with the tallest bearers at the head end of the coffin and the shortest at the front

▲ Six soldiers carry the coffin containing Lieutenant Corporal AR Smith of the 7th Battalion Australian Regiment at Rusthall in Kent in August 1916

◄ Having lifted the coffin from the trestles, the bearers move slowly over this uneven churchyard; they provide additional security for the coffin by grasping the handles

◄▼ The bearers walk with and then wear surplices over their black suits at the funeral in July 1909 of the Revd T H Bush, vicar of Christ Church in Dorset

◄ Four bearers carry the coffin at the funeral of a Scout. From the blackout on the hearse light, this image was probably taken during WWII

▲ The tradition of two bearers supporting a coffin is still maintained by staff at WG Miller in Islington.

LOWERING THE COFFIN

After the coffin has been transported to the place of burial or cremation, it then needs to be committed. Although the manual act of committal has largely remained unchanged for centuries, burial in places other than earth graves, in addition to dealing with large coffins and the use of mechanical devices, has altered the approach. From the perspective of the undertaker, the lowering of the coffin into the grave must be effected in a dignified manner; the objective is a slow and even descent.

In centuries past, when the coffin was only used as a means of conveyance to the place of burial, the body would be removed and handed or lowered down by rope into the grave. Burial of the body only, rather than within a coffin, is emphasised by the wording of the rubric from the 'Order for the Burial of the Dead' contained in the Book of Common Prayer (1662) where it states: 'When they come to the grave, whilst the body is made ready to be laid into the earth, the minister shall say...'

Burial of the body in a coffin alleviated the need for a person to stand in the grave as the coffin could simply be lowered using ropes or webs. This also enabled graves of considerable depth to be used. However, as these images show, there are several variations in how the coffin is prepared immediately prior to burial. In this image, a coffin is lowered into a boat, helped by the guiding hands of an anxious undertaker.

◄ In this image of the funeral of a priest, the coffin rests on trestles at the side of the grave before it is lowered. Wide-gauge webbing is clearly visible

▲ The coffin rests on putlogs over the grave. The word 'putlog' (also spelt 'putlock') is defined as 'a short horizontal beam that with others supports the floor planks of a scaffold'. The coffin is ready to be lowered, in this image of a naval funeral

◄ The coffin ready to be lowered at a Salvation Army funeral

THE NATIONAL BURIAL DEVICE

As used at the London Necropolis Cemetery.

For the Up-to-date Undertaker.

READ THIS.

One of the most distressing features of an English funeral is the lowering of the coffin. Nothing calls louder for reform! Anything you can do to alleviate your patrons' feelings in this respect will be appreciated by them. The National Lowering Device solves the problem. It lowers automatically at an even and reverent pace; does away with all struggling; and that which grieves your patrons most.

Write for full particulars and prices to the Sole Agents for the British Isles :—

DEVICE IN POSITION TO RECEIVE COFFIN.

The **"National"** is:

Easily Operated.

Convenient to handle.

Endorsed by Everybody.

Simple, Practicable, and Indispensable to the Up-to-date Funeral Director.

The **LONDON NECROPOLIS COMPANY,** 188, Westminster Bridge Rd., London, S.E.

▲ By the late nineteenth century, a number of ideas to improve the experience of mourners in the cemetery were arriving from America: the tent or awning to provide shelter against inclement weather; grass matting to cover the spoil; shoring apparatus to prevent collapse of a grave; and, finally, the coffin-lowering device. In June 1901, *TUJ* published an advertisement for 'The National Burial Device', which could be purchased from the sole agents in England, The London Necropolis Company. It stated, 'One of the most distressing features of an English funeral is the lowering of the coffin. Nothing calls louder for reform. Anything you can do to alleviate your patrons' feelings in this respect will be appreciated by them. The National Lowering Device solves the problem. It lowers automatically at an even and reverent pace; does away with all struggling, and that which grieves your patrons most.' It concluded by saying that the device was, 'Simple, practicable, and indispensable for the up-to-date funeral director.' In a later advertising feature it was noted that the idea was, '...American, of course...' and that the device could be used for children's coffins and for caskets up to 7½ ft in length. Both Brookwood and Kensal Green Cemeteries advertised the availability of a lowering device

GRAVE LOWERING DEVICE.

On Sale.

On Hire.

MANY advantages are to be obtained in using this "Knock-Down" Lowering Device. The size can be changed from 44 inches to 93 inches in length and from 20 inches to 34 inches in width. It can be placed in small space when ready for transportation. It is supplied with ball bearings and two brakes, which act independently. It can be taken from the cases in which it is carried, and placed ready for use in less than one minute. It is light, strong, neat, and easily fixed in position. When adjusted to a child's size it can be changed from 44 to 59 inches in length.

◄ Dottridge Bros also supplied the 'Knock Down' lowering device that was advertised in 1901. The notion of enhancing the experience at the cemetery re-emerged in 1928 when *TUJ* suggested that undertakers make representations to cemeteries in respect of '...three essential improvements...the clearance of the freshly dug earth away from the grave, an automatic lowering device, and a covering over the grave...' The lowering device should be '...part of the equipment of every burial authority.' Although in April 1936, Ingall, Parsons, Clive & Co enthused that, 'The steadily growing demand for this lowering device is proof that there is nothing more impressive than its use for the last solemn rite of burial', it would appear that the apparatus was not utilised to any great extent.

▲ One issue when lowering is the evenness of the descent. In 1908 Ingall, Parsons, Clive & Co became agents for the, 'Ward's patent safety Combination Cradle', a device with six webs that could be wrapped around the coffin. The advertising stated that, 'The coffin is held in such a position by the cradle that if any pair of bearers happens to lower quicker than the others, it is impossible for the coffin to slip.' But, like the lowering device, the cradle did not appear to be embraced by undertakers or burial authorities. Burial of the coffin other than in earthen graves has required different approaches to the committal. For example, small coffins for children are often handed to a member of the operational staff standing in a more shallow grave *(TUJ)*

▲ The coffin containing 28-year-old Flight Lieutenant Jack McPherson Richardson of the Royal Australian Air Force after being lowered into a grave at Oxford's Botley Cemetery in January 1945. Note the large inscription plate with the service number

▲ The burial of the ashes of Andrew Bonar Law in Westminster Abbey, October 1923. The development of cremation in the late nineteenth century posed the problem of how the casket of ashes should be buried. Should the casket be on display prior to burial or should it be covered or enclosed? When at Westminster Abbey it was decided that cremation must precede interment, the arrangement was to place the casket into a full-sized coffin, which was then covered by a hearse-cloth. At the moment of committal, a door in the corner of the coffin would be opened, the casket lifted out and then placed in the grave

OCCASIONAL TASKS

In addition to arranging the burial, cremation or onward transportation of the deceased, undertakers also carry out other tasks on an occasional basis.

CHURCH DRAPING BY JOHN W. HARLAND, Funeral Director.

▲ Undertakers were requested to drape the interior of houses and churches. This image shows the drapings arranged by John W Harland of Manchester. The records of J H Kenyon show that following the death of Queen Victoria in January 1901, they draped the banking premises of the London and County Bank Ltd on Edgware Road with purple velvet and stuff mourning, silk cords and tassels at a cost of £30. Bunney France of A. France & Sons in Holborn recalls undertaking this task in Roman Catholic churches up until the 1950s

▲ Although the preference for cremation during the early part of the twentieth century was only modest, one task that fell to the undertaker was the burial of caskets containing ashes. This image shows an undertaker holding a casket in front of the statute of Lord (William) George Frederick Cavendish-Scott-Bentinck (1802–1848) at Cavendish Square in London. It has not been possible to ascertain the circumstances surrounding this photograph

▲ An advertisement included on a local map for the Marylebone firm with the unusual name of Enefer

TELEPHONE 139 WILLESDEN.

GEORGE BURCH.

FUNERAL FURNISHER & MONUMENTAL MASON

158, Manor Park Road and

134, High St., Harlesden, N.W.10.

▲ This splendid Arts and Crafts advertisement for George Burch of Harlesden comes from a municipal guide dating from 1923

1162   *Army and Navy Co-operative Society, Ltd.*   Telephone No.: Victoria 850

FUNERALS AND CREMATIONS

THE Society has complete arrangements for the carrying out of **FUNERALS** and **CREMATIONS** upon any scale, in the best manner, and with special regard to economy.

Careful estimates will be submitted, on request, to meet particular requirements.

Full information regarding **CEMETERY** Charges can also be furnished, thus rendering the estimates entirely inclusive and free of all extras.

Facilities are available for the completion of **FUNERALS** in **PROVINCIAL TOWNS** and in the **COUNTRY**.

Well-equipped **MOTOR** Hearses and Carriages are in use and will be found most convenient for the conveyance of Remains to a distance, transferences en route being avoided.

FUNERALS can be conducted to and from places **ABROAD**, Shipments, Consular matters,

and Permits being dealt with by thoroughly experienced Managers, who are able to render the fullest assistance and to give expert advice on all points.

CREMATIONS.
—Pamphlets and Copies of the necessary Documents can be forwarded on application, together with all needful information.

ORDERS for Funerals or Cremations by **TELEGRAM** are promptly attended to at all times, irrespective of the ordinary business hours.

The Society's Registered Telegraphic Address is: "**ARMY, SOWEST, LONDON.**"

Instructions by **TELEPHONE** can be received and will be immediately dealt with. Telephone: "**VICTORIA 8500, EXTENSION 379.**"

After closing of Stores, and on Sundays and Holidays: "**VICTORIA 2978.**"

EMBALMING UNDERTAKEN.   **PORTRAIT SCULPTURE.**   **MONUMENTS.**

Valuations of real and personal property for Probate and other purposes.

COMBINED HAND-HEARSE AND BIER.

Designed to meet the requirements of Country Villages. Light in construction and strong.

In English Ash, stained and varnished, the ironwork painted black, with cycle wheels and rubber tyres, the Bier detachable and with hinged handles to drop **£32 10 0**

Do. do. without detachable Bier, light undercarriage, cycle wheels and rubber tyres, fold-back handle in front to draw   ..   **£18 10 0**

No. 8a.   (Carriage unpaid.)

◄ At the turn of the last century, many department stores operated a funeral service. Some examples from London include Harrods in Knightsbridge, William Whiteley at Bayswater and Ely's in Wimbledon. In many cases, the service was subcontracted to a local undertaker. This advertisement from their 1933 store directory is for the Army and Navy funeral service, located in their Victoria Street premises

▶ From the mid-nineteenth century, some undertakers produced literature describing their range of coffins and cost of funerals. In some cases, advertisements in local newspapers contained this information. In this page from a price list issued by J H Kenyon around 1900, the 'class' of funeral is based on the type of coffin and the degree of ostentation of the horse-drawn hearses

FIRST CLASS.

A Funeral Car or Hearse, drawn by four horses; three superior broughams, drawn by pairs of horses; an inner shell lined with flannel, edged with rich white satin; a fine flannel and satin burial robe or folding sheet; a stout oak outer case, French polished, or covered with black velvet or cloth, with massive brass handles of appropriate design; an engraved brass inscription plate; use of velvet pall; funeral service books for mourners; attendance of conductor, and the necessary assistants. £42

SECOND CLASS.

A Funeral Car or Hearse, drawn by four horses; two superior modern broughams, with pairs of horses; an inner shell lined with embossed cambric; a fine cambric folding sheet; a leaden coffin with inscription plate; a stout elm outer case, French polished, or covered with black cloth, with brass handles of an appropriate design; an engraved brass inscription plate; use of velvet pall; funeral service books for mourners; attendance of conductor, and the necessary assistants. £28

Or with oak outer coffin, French polished,
and finished with brass handles ... £32

THIRD CLASS.

A Funeral Car or Hearse, drawn by four horses; a very superior modern Carriage, drawn by pair of horses; an inner shell; an oak outer coffin, French polished, with brass handles of an appropriate pattern, and an engraved brass inscription plate; use of velvet pall; funeral service books for mourners; attendance of conductor and the necessary assistants. £21

FOURTH CLASS.

A Funeral Car or Hearse, drawn by pair of horses; a very superior modern Carriage and pair of horses; an inner shell, lined with embossed cambric; an elm outer coffin, French polished, with appropriate furniture and inscription plate; use of velvet pall; funeral service books for mourners; attendance of conductor and the necessary assistants
£12 12s. to £15 15s.
Or with leaden coffin £22

FIFTH CLASS.

A Funeral Car or Hearse, with pair of horses; a superior modern Carriage and pair of horses; stout elm coffin, French polished, with appropriate brass handles; an engraved brass inscription plate; use of velvet pall; funeral service books for mourners; attendance of conductor and the necessary assistants .. £10 10s.
Or with oak coffin ... ... £12 12s.

▶ As time progressed, the price list became more of a brochure that included images of the coffins. These two pages from a brochure issued by the Croydon firm of Thomas Ebbutt & Sons date from the 1940s. Named after an area of south London not far from Croydon, the 'Norwood' coffin was made of elm and then covered with white, purple or grey coloured cloth. The exterior was furnished with wooden handles. Such coffins were very popular for cremation and continued to be offered by funeral directors until the 1980s. Ebbutt's was founded in 1718 and at the time this brochure was produced, had five branches in the north Surrey area.

NORWOOD

NORWOOD
(CREMATION)

Coffin

Made of Elm, and covered with a choice of white, purple or grey cloth.

The interior is upholstered with swansdown and cotton wool and flannel and satin covering and pillow.

Furnished with Oak ring handles and engraved inscription plate.

Professional Services

These include attendance at residence if desired, to take instructions with regard to all arrangements for the entire funeral, conveying relatives to Registrar of Deaths, arranging with the Crematorium Authorities and the appropriate Minister to officiate.

A fully qualified assistant's attendance to carry out the preservation of the remains.
Payment by us (and added to our account) on your behalf, all expenses including Crematorium and Doctor's fees for Cremation Certificates, Church fees, Minister's fee, all notices in the daily or weekly press or other publications as desired, and any other special instructions we are requested to carry out.

Ancillary Services

Conveyance and attendants delivering coffin and placing remains therein.
Providing a stand with velvet pall on which to place coffin.
A Rolls Royce motor hearse and chauffeur and a Rolls Royce car and chauffeur for mourners.
Five bearers' attendance at funeral and their conveyance.
Providing a revolving top catafalque draped with velvet pall in Church.
Superintendent to take charge of all arrangements at the time of funeral, and Personal Superintendence throughout.
For the sum of £43 18s. 10d.

▲ A highly decorative letterhead for J Marshall, building contractor and undertaker of Buckingham. Capable of making mantlepieces and stoves, their construction skills extended to the erection of glasshouses and small buildings. In 1906, the image for their 'completely furnished funerals' was quite dated

▼ After the funeral, an invoice would be submitted to the family. Many firms had elaborate letterheads that itemised their range of service and often included illustrations. This one dating from 1905 for Francis Chappell not only lists all their branches, but also indicates that they could supply hatchments. The compass and dividers and the beehive denotes that they have links with freemasonry

TELEPHONE:
CLERKENWELL 3152.
„ 3790.

27, FEATHERSTONE STREET, CITY ROAD,
London, E.C.

Branch
III, CENTRAL ST
ST LUKES.

19

M Mr Martin

Feb 6 1934

Dr to H. E. Pierce,
THE CITY
Funeral Furnishing Warehouse.

MOTOR HEARSES, LANDAULETTES
AND FUNERAL CARRIAGES OF
THE BEST DESCRIPTION.

Established
1840.

FUNERAL CARRIAGE PROPRIETOR.

Lead, Oak & Elm Coffins. Monuments & Tombs erected.

▲ H E Pierce's main office was on Featherstone Street; they also had a branch on the City Road. Established in 1840, when this invoice was submitted just under a century later, the firm was still providing horse-drawn funerals, although a concession had been made to motor vehicles

▶ Frederick W Paine had a large masonry works at Fairfield West in Kingston and a line drawing was depicted on their invoices. The pillared way to the left of the centre provided access to the firm's chapel of rest.

Chief Office Telephones:
KINGSTON 0828 } TWO LINES
1796

FREDᵏ W. PAINE
Monumental Sculptor and Mason

Works, Studio & Show Rooms. 23, 24, 25, 26, 27, 29, 30 & 31, Fairfield West.
CHIEF OFFICE:
24, LONDON ROAD. **KINGSTON-on-THAMES.**
ALSO AT TEDDINGTON,
TWICKENHAM,
AND NEW MALDEN.

▼ Dating from 1928, this letterhead from C Selby & Son includes a burning torch, an emblem of enlightenment and hope.

Telephone: 3034 EASTERN.
AND
2064 MARYLAND.

London, 25th August. 192 8

Mrs. Watling, 35, Franklin Street.

HEAD & FOOT
STONES ERECTED
AT THE
LOWEST POSSIBLE CHARGES
CARRIAGE DEPARTMENT
ARROW ROAD

TO **C. Selby & Son,**
J.W. SELBY
FURNISHING UNDERTAKERS
AND **FUNERAL CARRIAGE PROPRIETORS.**

31, CAMPBELL ROAD, E.
191, HIGH STREET, STRATFORD, E.
AND AT
729, HIGH ROAD, LEYTONSTONE, E.

146, BOW ROAD, E.3.

▲ The Chiswick-based funeral furnishers and monumental masons W.G. Barratt was founded in 1781.

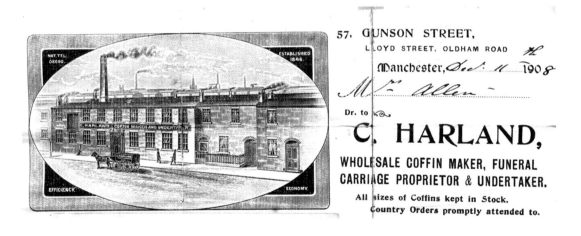

▲ Established in 1846, the Manchester-based firm of C Harland made coffins but were also undertakers. As can be seen on page 269, they also draped the interior of churches.

MONUMENTAL MASONRY

Up until the 1950s, the vast majority of funerals in Britain were burials. For this reason, it is understandable that there was a close relationship between undertaker and monumental mason. Whilst some firms simply recommended a mason (for which they received a commission) and had a display in their window (as seen in the image of J Hawes' premises on page 203), others operated a fully equipped masonry department. Space was always needed for storage and to work on the stones, in addition to a display area for potential clients. Many deemed it advantageous to be situated near or opposite the entrance to a cemetery, not only for commercial reasons but also for ease of transporting new memorials.

▶ A view of T Candy's masonry works at Lewisham. The business had branches at Elmers End Road and Croydon and was ideally placed to serve the cemeteries at Lewisham and Lee, as the advertisement states

▶ The Art Memorial Co (established by the late W Piper in 1837) was located opposite the entrance to West Norwood Cemetery in south London. The masonry business has disappeared, but the stone building in the background is now part of Yeatman's funeral directors

▲ The display area at Jordan's masonry business on Acton High Street was the front garden of a large Victorian house. The second image shows a tram passing the firm's works situated behind a large advertising hoarding

◄ White marble memorials with angels and crosses in an unidentified mason's yard, awaiting erection in a cemetery, or the addition of a new inscription

▼ Masons at work on the entrance to Hollybrook Cemetery at Southampton. The cemetery opened in 1913

FRANCIS & C. WALTERS, LTD

Undertakers and Monumental Masons,

TELEPHONE :- EAST 1289.   Estimates Free upon application.

811, COMMERCIAL ROAD, LIMEHOUSE, E.14.

ALSO

87, UPPER NORTH ST., POPLAR & 15, HIGH ST., STEPNEY.

▲ Designs from a mason's catalogue probably dating from around 1910

▶ Described in the brochure as 'an effective memorial', this marble angel on the grave of Lily (aged 22) and Samuel (aged 30, who was buried elsewhere) dates from around the time of WWI. It is surrounded by graves marked by Immortelles (see page 284)

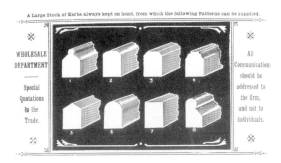

A Large Stock of Kerbs always kept on hand, from which the following Patterns can be supplied.

WHOLESALE DEPARTMENT

Special Quotations to the Trade.

All Communication should be addressed to the firm, and not to individuals.

▲ Designs for kerbs from a brochure issued by the Artistic Monumental Co of London around 1913

EXPORT.

Packing in cases and delivery to docks (London) 5 per cent. extra on cost.

Freight and Insurance F.P.A. charged at cost and which is usually covered by a further $7\frac{1}{2}$ to 15 per cent. dependent upon values and weight.

Inclusive Prices quoted delivered to any Port in the World.

Scroll and Pillars in Marble, height 2 ft. 6 in., with Curbing on Stone Plinth, £25.

No. 02120   Erected at ISLINGTON.   *Copyright.*

▼ Designs from a mason's catalogue published in the mid-1920s

Ledger in Silver Grey Granite, on Landing, **£35.**

No. 0675   Erected at KENSAL GREEN.   *Copyright.*

Ledger Tomb in Marble, on Landing, 6 ft. 6 in. by 2 ft. 6 in., **£57.**

No. 0559   Erected at BROOKWOOD.   *Copyright.*

J. UNDERWOOD & SON, Ltd., BAKER ST., LONDON.

Erected at **WANDSWORTH.**

No. 1822   Copyright

Rifle Cross in marble, height 5' 6".
Price - - **£42 0 0**   Curbing extra.

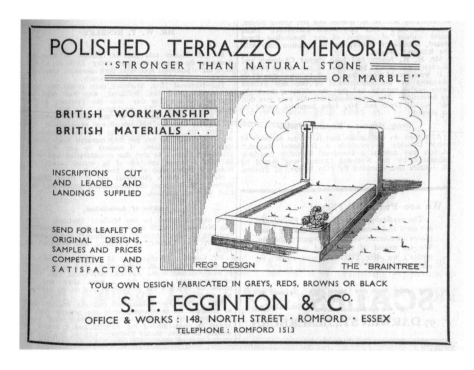

▲▼ Two designs from the 1940s and 1950s of memorials made from artificial stone *(BUA Monthly)*

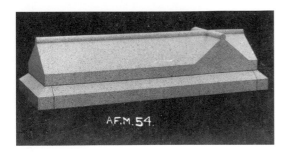

▲ Two Cornish granite designs from 1940 supplied by the Ashford based mason, F H Russell

◄ (Left) William Chowing was an Exeter monumental mason. This image shows the memorial erected on his grave in1905

◄ (Right) A memorial erected in 1913 by A. Jacobs monumental masons, whose works were conveniently situated opposite Tottenham Cemetery

◄ Once the memorial had been placed at the graveside, the family would visit. This scene shows a laurel wreath about to be placed on the grave. An Immortelle is clearly visible

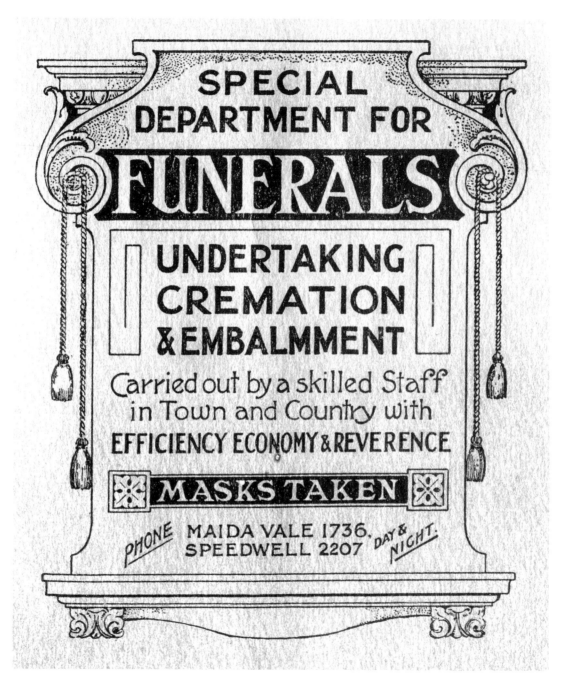

▶ The firm of John Underwood, formerly of 27 Baker Street, relocated to 38a Boundary Road St John's Wood in the 1930s. They were monumental masons but also had a funeral department. As can be seen from this advertisement in their masonry catalogue, Underwood's could provide a death mask.

OUT ON FUNERALS

A hand-pulled gun carriage was used to convey Lieutenant Loftus Masaragh's coffin to Clayhall Cemetery in March 1904. He was one of 11 to die when an A1 submarine was struck on the starboard side by SS *Berwick Castle* near Portsmouth Harbour. These images show the funeral procession and the scene in the cemetery.

One of the first hit-and-run accidents ever to take place in this country was in the Hertfordshire village of Markyate, when four-year-old Willie Clifton was knocked down and killed outside his home on London Road in April 1905. The incident aroused much local and national attention as the proprietor of the *Daily Mail*, Sir Alfred Harmsworth, offered a reward of £100 for information about the driver of the vehicle. It was discovered, however, that the driver was Rocco Cornalbas, chauffeur to Hildebrand Harmsworth, Sir Alfred's brother. Although Cornalbas was alone in the car, he was said to be driving at a 'terrific pace', but did not think anything serious had happened when the car struck young Willie. Cornalbas was tried at the Hertfordshire Assizes in July 1905 and, after being found guilty of manslaughter, was sentenced to six months' hard labour. Willie Clifton's funeral took place on 22 April and this image shows the open coffin with the coffin lid bearing an elaborate nameplate standing against a wall. The other image shows the scene around the grave in the extension to the churchyard

The funeral of conductor Walter Robinson who was killed in October 1907 when the brakes of a Halifax Corporation tram failed and it came off the rails at Pye Nest Road, Halifax. A single-deck tram was used at his funeral to convey his coffin.

The coffin of Lord Nunburnholme surrounded by navy personnel is carried for burial on 31 October 1907. Charles Henry Wilson was born in April 1833 and joined the family business of Thomas Wilson & Sons, ship owners. By 1891 they had a fleet of 100 vessels and was considered the largest such private company in Britain. Charles Wilson served as High Sheriff of Hull and was elected a Liberal MP in 1894. Given the Freedom of Hull in 1899, he was raised to the peerage as Baron Nunburnholme of the City of Kingston upon Hull seven years later. He died on 27 October 1907

The funeral of Sir Eyre Massey Shaw, superintendent of the Metropolitan Fire Brigade and its predecessor, the London Fire Engine Establishment, from 1861 to 1891. He was responsible for introducing modern fire-fighting methods and the famous brass helmets, along with increasing the number of fire stations. He died at Folkstone on 25 August 1908. Here the hearse is seen entering the gates of the eastern part of Highgate Cemetery

The funeral held on 18 November 1908 of Vice-Admiral Sir Henry Deacon Barry KCVO. The coffin was taken by train from Chatham to Botley in Hampshire, where it was placed on a gun carriage drawn by men of the *Excellent*. Among the pall bearers were Admiral Sir Arthur D Fanshawe, Commander-in-chief at Portsmouth and Vice-Admiral Sir Francis C B Bridgeman, Commander-in-chief of the Home Fleet. The funeral procession comprised military representative of the Royal Marine Light Infantry and the Royal Marine Artillery, along with officers of the dockyard, civil officers, army officers, naval officers and the rank and file. The service was conducted by the rector of Botley and the chaplain of the *Indomitable*, who had served as chaplain on the Vice-Admiral's flagship.

A gun carriage is used to convey the coffin containing Lieutenant Roy Maurice Gzowski in September 1910. Arriving in England only a few months prior with the 2nd Queen's Own Rifles of Canada, he died from enteric fever and was buried at Aldershot Military Cemetery. The memorial over his grave was designed with the assistance of Mrs G F Watts of the Potters' Art Guild at Compton

The funeral at St Paul's Cathedral of three policemen shot during a raid of a jewellery business in Houndsditch in December 1910. The incident led to the infamous 'Siege of Sydney Street'. Sergeants Bentley and Tucker and Constable Choate were posthumously awarded the King's Police Medal; the two sergeants were buried in the City of London Cemetery at Ilford, and Constable Choate at Byfleet in Surrey

A fire in the premises of Messrs A Cohen, rag merchants, on Bankside in south London resulted in the death of five firemen when they were trapped under a collapsed staircase. The funeral of Fireman Willan took place at Southwark Cathedral on 6 March 1911, followed by burial at Woodgrange Park Cemetery in east London

Ordained in 1868, the Venerable William Henry Askwith was appointed vicar of St Mary's Taunton in 1887. He became Archdeacon of Taunton in 1903. His death on 9 April 1911 was followed four days later by his funeral. This image shows the procession entering the church

THE GREAT LAYFAYETTE

Born in Germany, Sigismund Neuberger was known to enthralled music hall audiences as 'The Great Lafayette'. His status as one of the world's greatest illusionists enabled him to command around £500 per week. Lafayette was besotted with his dog, Beauty, who had been given to him by Harry Houdini. Beauty enjoyed five-course meals, wore a jewel-encrusted precious metal bracelet and slept in a gold-plated bedstead at Lafayette's house in Tavistock Square in London.

During 1911, Lafayette was engaged on a tour of Britain that included two weeks in May at the Empire Theatre in Edinburgh. Tragically, however, Beauty died on Saturday 6 May. The inconsolable Lafayette made arrangements for the dog to be embalmed and then buried in a vault lined with white enamelled tiles in Edinburgh's Piershill Cemetery. Beauty's funeral was scheduled for Wednesday 10 May; in the meantime, the dog lay in state in a glass case in the city's Caledonian Hotel. The cemetery had agreed to bury Beauty provided that Lafayette was subsequently buried in the same grave; ironically, this would be sooner than anticipated.

On Tuesday evening, Lafayette and his troupe entertained a capacity audience to a show culminating in a performance of 'The Lion's Bride'. Dressed in oriental costume, he appeared on a horse amid a scene of women in a harem. Against a backdrop of lights and exotic scenery, a young girl was captured and thrown into a lion's den with the real lion. The audience rightly gasped at this spectacle. Having changed placed with a double, Layfayette entered the cage dressed as a lion while the real lion exited the cage using a trap door. He then revealed his identity. However, an electric wire above the stage had fused and the resulting fire rapidly spread throughout the stage area. After the fire had

Funeral of The Great Lafayette, Entering Piershill Cemetery.

been extinguished, Lafayette was discovered lying on the stage floor; his identity was confirmed by Lafayette's manager and several members of the company. His body was then taken by train to Glasgow for cremation and his ashes scheduled to be buried at Piershill on Sunday 14 May. *The Times* reveals what happened next:

> *'While search was being made among the debris beneath the stage for the missing body of Charles Richards, who acted as his [Lafayette's] "double" in the performance, the searchers came across a body. It was, of course, assumed at first to be that of Richards, but on the fingers of the body were found to be a gold ring and a diamond ring, and it was at once realized that after all this was the body of "Lafayette", and that it was the body of Richards that had been cremated in mistake for his.'*

Referring to the sword and costume that had been the chief distinguishing 'marks' on the body believed to be that of Lafayette, *The Times* noted:

> *'The presence of these is now explained by the fact that the body was that of the man who acted as his "double" in the sketch which he was performing at the time of the fire, and who thus in a tragic manner kept up the "illusion" even after death.'*

The real Lafayette's body had been discovered in a basement of the stage. He was in a cramped position and his features charred. His solicitor, who had travelled from London, had no difficulty in confirming Lafayette's identity, as did several members of his troupe. He was taken by train to Glasgow where he was cremated and his ashes returned to Edinburgh. Despite the tragedy, but as planned, Beauty had been buried the day after the fire.

On Sunday 14 May, the cortege comprising twenty carriages passed the Caledonian Hotel, down Princess Street towards Piershill Cemetery. The hearse carried Lafayette's ashes in a lead-lined oak casket; the inscription read, 'The Great Lafayette, who perished in the Empire Palace Fire, May 9, 1911.' The military band of the Edinburgh branch of the Amalgamated Musicians' Union headed the procession. As the local rabbi had refused to conduct the funeral as the interment was in unconsecrated ground, an Episcopalian curate read the burial service. Lafayette's casket was placed between the front paws of Beauty and the glass case was replaced in the vault. An inscription in the lower right hand corner of the memorial simply records 'The Great Lafayette'.

The funeral in March 1911 of the Hon Percy Wyndham at East Knoyle in Wiltshire. Born in 1835, he was the son of the first Lord Leconfield. A former Conservative MP for West Cumberland, Wyndham was also Deputy Lieutenant for Cumberland, Sussex and Wiltshire. His funeral took place at East Knoyle on 18 March. The London and South Western Railway Company ran a special train from Salisbury to Semley station to convey those from London wishing to attend the service.

The first image shows the coffin on a covered bier in the church; the second shows the bier with the coffin carried shoulder-high through the rain.

The funeral of the pioneering aviator Col Samuel F Cody, who died in August 1913. Killed whilst on a test flight when his Cody Floatplane broke up at 500ft, he was given a funeral with full military honours. The photos show the coffin being taken from his house in Ash Vale, Surrey, and placed on a field-gun carriage drawn by a team of six horses ridden by scarlet-uniformed engineers. The two-mile procession to Aldershot Military Cemetery was lined by an estimated 100,000 people. The floral tributes ranged from a 6ft-high propeller to Mrs Cody's steering wheel coffin wreath

The funeral on 5 September 1913 of victims killed in an accident on the Settle to Carlisle railway line. In the early hours of 2 September, a train stalled on a lonely part of the line about half a mile short of the summit at Aisgill. A second train then passed red signals on the line and collided with the first train. Of the funeral, *The Mid Cumberland & North Westmorland Herald* said, 'The inhabitants will never forget the remarkable scene as the bearers emerged from the church premises and passed in one long line along Market Street, carrying their mournful burdens shoulder-high. No, they will never forget the sorrowful spectacle, as one after another, the nine coffins arrived at the place of interment and were reverently lowered into their respective graves.'

An Autocar being used at a funeral during WWI. This smallish American truck had been converted into a lightly armoured machine-gun carrier. Only 20 of them were built, and were operated by the First Canadian Motor Machine Gun Brigade. This vehicle left Canada in September 1914 under the command of a French Canadian officer, Major Raymond Brutinel, and spent eight months in England, mostly at Shorncliffe, before going across to France. The location of the funeral is believed to be in the Folkstone area

A firing party at the funeral of the Revd Clement Larcom Burrows, vicar of St Paul's, Bournemouth and eldest surviving son of Major-General Burrows RA. He died on 17 February 1915 and was buried three days later

These two buses festooned with flowers formed part of the procession at the funeral of driver Charles James Tarrant and conductor Charles Rogers who were killed in October 1915 when their bus was blown up during an air raid. Over 700 colleagues together with officials of the London Omnibus Co, members of the London and Provincial Union of Licensed Vehicle Workers, firemen, soldiers and policemen followed the buses from Cricklewood Garage to Willesden Cemetery

FUNERAL OF VICTIMS OF AIR RAID. DRIVER TARRANT AND CONDUCTOR ROGERS. L.G.O.C. 20/10/15. E.S. 3.

FUNERAL OF VICTIMS OF AIR RAID. DRIVER TARRANT AND CONDUCTOR ROGERS. L.G.O.C. 20/10/15. E.S. 7.

Following her exhumation from the Tir National Cemetery in March 1919, just three years and five months after being executed in Brussels for helping soldiers escape, Nurse Cavell's body was brought by train from Dover Marine station to Victoria and then by gun carriage to Westminster Abbey. After the service, her coffin was taken through the city to Liverpool Street station for the final journey to Norwich where she was buried with military honours at Life's Green in the shadow of the cathedral

▲ Edith Cavell (1865–1915)

▶ King George V and Queen Mary visiting her grave

▶ The exhumed body

▼ Nurse Cavell's execution was used as propaganda by the British government. This is one of the many postcards issued

▶ Carrying the coffin from Tir
National Cemetery

▶ The procession down Queen Victoria
Street towards Liverpool Street station.
Arthur Whitehead, an undertaker based in
Rochester Row, Westminster assisted with the
arrangements for the funeral in London

▶ Nurse Cavell's grave at Life's Green

THE R38

At 695ft long and 85.5ft wide, the R38 was at the time the largest airship ever to have been built. Designed by the Admiralty, she was constructed at the Royal Aircraft Works at Cardington in Bedfordshire. She made her first trial flight on the night of 23 June 1921, leaving Howden airship base in the East Riding at 7.10 a.m. destined for Pulham in Norfolk. She remained out overnight to complete tests, which were reported to be satisfactory. The next afternoon, staff signalled that due to low cloud and difficulties with mooring, she would return to Howden around early evening.

At 5.30 p.m. the airship was flying at about 1000ft and some high-speed turns were carried out over the Humber. Suddenly she took a nosedive and broke into two sections; this was followed by two terrific explosions. The front part of the airship descended onto a sandbank in the middle of the river, which was at low tide, while the rear floated down to the other side of the river. Thousands of people made their way to the pier to see the wreckage; many had been alerted to the accident by the explosion, which had shattered windows in the older part of the city. Rescue tugs were hampered by the low tide. *The Times* quoted one eyewitness:

> *'...when the airship touched the water one half exploded, and a few seconds later, with a terrific roar, the second half exploded. When the material touched the water a huge mass of flame and smoke was all that could be seen. When the smoke cleared away all that could be seen was about half of the airship projecting above the river.'*

An Admiralty representative visited Hull and issued this statement:

> *'The wreck appears to be in two portions. The whole vessel lies submerged in a north-west direction in the river in eight feet of water, the tail portion only showing and lying in about four feet of water at half tide. The survivors escaped by parachute, and were picked up by tugs and small boats. The Custom House has been ordered to guard the wreck during the night, and an Air Force officer from Howden will patrol the river in a motor-launch to look out for bodies and wreckage which may drift into the river.'*

A news release stated that only five of the total complement of 49 American and British on board were known to have survived. The casualties included Air Commodore Edward Maitland of the RAF and the US Navy's Commander Maxfield. Flight Lieutenant Wann in command of the R38 survived, but was 'gravely hurt', including severe burns.

By Friday, only three bodies had been found; two more were recovered on Saturday. It was suspected that many of the dead lay beneath the wreckage and an 80-ton floating crane was utilised for the salvage operation. By Sunday, considerable quantities of the frame, gas bags and parts of the engine had been recovered. In total 44 people died; 16 were US servicemen and 28 were British

An RAF tender drawing the aircraft trailer containing some of the coffins and floral tributes.

On Wednesday 31 August, the announcement was made that the funeral of five of the British officers and servicemen would take place on Friday 2 September. The procession started from Hull mortuary where the bodies had been placed in polished oak coffins with plain brass mountings. The nameplate included the name, rank, date of death (24 August 1921) and the age. Motor tenders drawing aircraft trailers served as biers to transport the coffins on the two-mile journey to Hull's Western Cemetery. The procession was headed by 100 men from Howden base with reversed arms, followed by the RAF Central Band playing Chopin's 'Marche funèbre'. Tens of thousands of people lined the streets to pay their respects. The chaplain-in-chief of the RAF, Revd H D L Viener and Revd W T Rees, chaplain at RAF Howden, conducted the brief service at the graveside. After a short silence, shots from the firing party rang out, with three volleys fired into the air. The Last Post concluded the ceremony.

On 5 September, *The Times* reported that the last American body had been recovered from the wreckage of the R38. The next day, they were embalmed in the mortuary at Hull Royal Infirmary. *TUJ* noted that:

> '...*two foremost American embalmers, Supervisor FG Kelly and JF Nadine of the American Grave Registration Service based in Paris arrived in Hull on August 29, in response to cabled instruction from Washington to embalm the bodies recovered from the ill-fated airship ZR2.*'

However, in the following month's edition, the *BUA Monthly* revealed that the bodies had already been embalmed:

> '*It will have been noticed that the United States Government decided that the recovered American bodies should be embalmed prior to being brought home, and to that end instructed two of the official embalmers to proceed from Paris to Hull to carry out the operation. Prior to the arrival of the American embalmers, however, Messrs Shepherd and Robinson, two of the Hull [BUA] Centre members, had been instructed to embalm the American bodies. The American embalmers came from Paris where they are engaged in caring for their own dead countrymen. Their orders were cabled from Washington and the local authorities knew nothing of the arrangement*

until these gentlemen arrived. The work carried out by Messrs Shepherd and Robinson, with the assistance of Mr Harry Andrew of Hull, was to the entire satisfaction of the American colleagues, who congratulated them on their thoroughness and efficiency.

As regards the funeral arrangements we learn that Mr Shepherd was entrusted with those of the Americans and Mr Robinson of the English. The caskets, sixteen in all, were sent by the American authorities at Southampton to Mr Shepherd, who then supervised the conveyance, coffining, lying in state, and the arrangements for the removal of the whole number. The American Air Consul was most warm in his praise of Mr Shepherd's work, for every item passed off without the slightest hitch of any kind.'

The bodies destined for the US were placed in metal-lined oaken caskets; inside were their uniforms and an American flag. Each casket was covered by the Stars and Stripes and also wreaths. On 6 September, the caskets were taken by train via London to Devonport dockyard at Plymouth. Departing from Hull at 6.30 p.m., they arrived at 10 p.m. Accompanied by an 18-man RAF guard of honour, they were met at Devonport by a naval guard to be placed on HM light cruiser *Dauntless*, which had been placed at the disposal of the American government to convey the coffins back to the United States. The journey commenced on the morning of 7 September. All were interred in America with the exception of Lieutenant-Commander Coll who, at the request of his wife, was buried at sea in American waters

The funeral of Algernon Seymour, the fifteenth Duke of Somerset, who died on 22 October 1923. His coffin was made from wood grown on his estate at Maiden Bradley near Warminster in Wiltshire, and the funeral was held in the village church, before an estate wagon transported the duke's coffin to the place of burial

One funeral that can almost claim to have utilised the London Underground system was for Dr John Barnardo, the founder of the children's charity known today as Barnardo's. At the time of his death in 1905, the line to Barkingside was served by Great Eastern Railway Company steam trains from Liverpool Street station: it was not until May 1948 before the station was incorporated into the system and underground trains first stopped at Barkingside station.

After a lying in state at the Edinburgh Castle Mission Hall and procession from Limehouse through the East End, Dr Barnardo's coffin was loaded onto a train at Liverpool Street. At Barkingside, six bearers carried the coffin the short distance to the Children's Church in the grounds of the Girls' Village Home. However, after a service in the church, his coffin was taken all the way to Woking for cremation and it is only his ashes that are buried under the huge memorial designed by George Frampton RA

The funeral of Father Andrew, one of the three founders of the religious Society of Divine Compassion and author of devotional books. Born Henry Ernest Hardy in 1869, he studied at Oxford and was ordained in 1895. The Society was based at the mission church of St Philip's in Plaistow. The building was demolished during WWII and Father Andrew died shortly after in March 1946. As this image shows, a long procession made its way through the streets of the parish in which he had worked for over fifty years, before reaching the East London Cemetery where he was buried in the St Philip's Guild Ground

PRIEST'S FUNERAL
Clergy wearing cottas and birettas line up outside this unidentified church at the funeral of a fellow priest.

With white handkerchiefs covering their faces, what is likely to be the principal mourners join with a large queue of people to follow this flower adorned coffin that rests outside a house on low trestles. The person on the right of the coffin is probably the undertaker.

APPENDIX
BOOKS, NEWSPAPERS AND PERIODICALS

Barclay R (2008)
We're Down Lads. The Tragedy of the Airship R101
Cardington: St Mary's Church

Barnfield P (2007)
When the Bombs Fell. Twickenham, Teddington and the Hamptons under Aerial Bombardment during the Second World War
Twickenham: Borough of Twickenham Local History Society

Bourke J (1996)
Dismembering the Male: Men's Bodies, Britain and the Great War
London: Reaktion Books

British Undertakers' Association Year Book and Diary 1919

Bradley DL (1986)
LSWR Locomotives: The Drummond Classes Didcot: Wild Swan Publications

Cannadine D (1981)
'War and Death, Grief and Mourning in Modern Britain' in J Whaley ed *Mirrors of Mortality*
London: Europa

Carr-Saunders AM, Sargant Florence P and Peers R (1938)
Consumers' Co-operation in Great Britain. An Examination of the British Co-operative Movement
London: George Allen & Unwin Ltd

Child S (1997)
'The Horse in the City' *The Victorian Society Annual 1996* pp5-14

Coates T ed (2001)
Tragic Journeys
London: The Stationery Office

Faulkner JN and Williams RA (1988)
The LSWR in the 20th Century
Newton Abbot: David & Charles

Fisher P (2009)
'Houses for the Dead: The provision of Mortuaries

in London, 1843-1889' *The London Journal* Vol 34 No 1 March pp1-15

Gavaghan M (2006)
The Story of the British Unknown Warrior
Fourth Edition Le Touquet: M&L Publications

Gittings C (1984)
Death, Burial and the Individual in Early Modern England
London: Croom Helm

Gordon WJ (1893)
The Horse World of London
London: The Leisure Hour Library

Hall S (1990)
Railway Detectives: 150 Years of the Railway Inspectorate
London: Ian Allan

Harkin T (2010)
Coventry 14th/15th November 1940. Casualties, Awards and Accounts
Coventry: War Memorial Park Publications

Hart V & Marshall L (1983)
Wartime Camden
London: Camden

Haslam MJ (1982)
The Chilwell Story
Nottingham: RAOC Corps

Hill G and Bloch H (2003)
The Silvertown Explosion: London 1917
Stroud: Tempus Publishing

Honigsbaum M (2009)
Living with Enza. The forgotten Story of Britain and the great flu pandemic of 1918
London: Macmillan

Irion PE (1970)
The Funeral and the Mourners: Vestige or Value
Nashville. Abingdon Press

Jalland P (1999)
'Victorian Death and its Decline: 1850-1918' in
Jupp PC and Gittings eds *Death in England: An
Illustrated History*
Manchester: Manchester University Press

Jalland P (2010)
*Death in War and Peace. The History of Loss & Grief
in England, 1914-1970*
Oxford: Oxford University Press

James S and Sargent T (2007)
'The Economic Impact of an Influenza Pandemic'
Department of Finance, Canada

Jupp P (2005)
*From Dust to Ashes: Cremation and The British Way
of Death*
Basingstoke: Palgrave MacMillan

Lawrence CE (1917)
'The Abolition of Death' *The Fortnightly Review* Vol
DCII 1 February pp326-331

Lewis T (1990) *Moonlight Sonata: The Coventry
Blitz 14/15 November 1940*
Coventry: T Lewis & Coventry City Council

Longworth P (1985)
The Unending Vigil: The History of the
Commonwealth War Graves Commission Barnsley:
Leo Cooper

Litten JWS (2002)
*The English Way of Death: The Common Funeral
Since 1450*
London: Robert Hale

Masefield PR (1982)
To Ride the Storm. The Story of the Airship R101
London: William Kimber

Morley J (1971)
Death, Heaven and the Victorians
London: Studio Vista

Moynihan M ed (1983)
God On Our Side
London: Leo Cooper/Secker and Warburg

Newberry CA (2006)
*Wartime St Pancras: A London Borough
Defends Itself*
London: Camden History Society

Nock OS (1987)
Historic Railway Disasters
Fourth Edition London: Ian Allan

Oddie S I (1941)
Inquest
London: Hutchinson

Parsons B (2004)
'Farewell to the Appendages of Sorrow: The End
of the Funereal Plume' *BIFD Journal* Vol 18 No
3 pp13-15

Parsons B (2005)
JH Kenyon: The First 125 Years
Worthing: FSJ Communications

Parsons B (2005)
Halford Mills: Funeral Reformer and Pioneer of
Embalming *FSJ* June pp64–72

Parsons B (2009) 'Unknown Undertaking: The
History of Dottridge Bros Wholesale Manufacturers
to the Funeral Trade' *Archive: The Quarterly Journal
for British Industrial and Transport History* No
63 pp29-41

Parsons B (2011)
'The Civilian War Grave in S Alban's Burial Ground
at Brookwood Cemetery' *Necropolis News* Vol 5 No
4 pp3-8

Pattenden N (2001)
Salisbury 1906: An Answer to the Enigma
Swindon: The South Western Circle

Pond C (2002)
Lying in State
House of Commons Library (SN/PC/1735)

Ponsonby A (1980)
'The Corpse Factory' *The Journal for Historical
Research* Vol 2 No 1 p121-130

Puckle BS (1926)
Funeral Customs: Their Origin and Development
London: Werner Laurie

Reader D (2008)
A Pictorial History of the British Hearse 1800-2008
privately published

Roffey R (1999)
The Co-operative Way
CWS: SE Region

Rolt LTC (1982)
Red for Danger
Fourth Edition, Newton Abbot: David & Charles

Rose A (2007)
Lethal Witness. Sir Bernard Spilsbury,
Honorary Pathologist
Kent (Ohio): Kent State University Press

Rugg J (2004)
'Managing 'Civilian Death due to War Operation':
Yorkshire Experiences during World War II
Twentieth Century British History
Vol 15 No 2 pp152-173

Rugg J (2005)
'Managing 'Civilian Death due to War Operations':
The Lessons of World War II' *The Journal of*
the Institute of Cemetery and Crematorium
Management Vol 73 No 1 pp40-42

Taylor L (1983)
Mourning Dress: A Social and Costume History
London: George Allen & Unwin

Walmsley N Le N (2001)
R101: A Pictorial History
Stroud: Sutton Publishing

Wensley FP (1931)
Detective Days: The Record of Forty-two years'
Service in the Criminal Investigation Department
London: Cassell

West J (1988)
Jack West: Funeral Director, 60 Years with Funerals
Ilfracombe: AH Stockwell

Wilkinson J (2006)
The Unknown Warrior and the Field
of Remembrance
London: JW Publications

Newspapers & Periodicals:

The Salisbury and Winchester Journal

The Times

The Liverpool Echo

The Stratford Express

The Willesden Chronicle

The Tottenham and Edmonton Weekly Herald

Croydon Times

The Daily Telegraph

The Embalmer

The Illustrated London News

Stanmore & Edgware Observer

The Scotsman

The Manchester Guardian

The Orcadian

Flight

The Daily Mirror

Sunday Pictorial

The National Funeral Director

The Undertakers' Journal

The Undertakers' and Funeral Directors' Journal

The BUA Monthly

The Producer

Co-operative News

The Scottish Co-operator

Co-operative Review

The Scottish Co-operator

The Co-operative Review

Comradeship and the Wheatsheaf